THE 'POEMS' OF

MRS. ANNE BRADSTREET
(1612–1672)

TOGETHER WITH HER
PROSE REMAINS

WITH AN INTRODUCTION BY
CHARLES ELIOT NORTON

THE DUODECIMOS
MDCCCXCVII

Copyright, 1897, by The Duodecimos.

THE DE VINNE PRESS.

CONTENTS.

	Page
INTRODUCTION, by Charles Eliot Norton	vii
EDITOR'S NOTE	xxxiii

THE POEMS OF MRS. BRADSTREET:

Prefatory verses by admirers	3–14
To her most honored father	15
The Prologue	17
The Four Elements	19
The Four Humors	37
The Four Ages	60
The Four Seasons	77
The Four Monarchies	87
A Dialogue between Old England and New	218
An Elegy upon Sir Philip Sidney	230
In Honor of Du Bartas	234
In Honor of Queen Elizabeth	238
David's Lamentation	243
To the Memory of My Father	245
An Epitaph on My Mother	248
Contemplations	249
The Flesh and the Spirit	259
The Vanity of all Worldly Things	263
The Author to her Book	266

Contents

	Page
Poems upon divers occasions:	
Upon a fit of sickness	267
Upon some distemper of body	268
Before the birth of one of her children	269
To my dear and loving husband	270
A letter to her husband	271
Another	272
Another	273
To her father, with some verses	275
In reference to her children	275
In memory of Elizabeth Bradstreet	279
In memory of Anne Bradstreet	280
On Simon Bradstreet	281
To the memory of Mercy Bradstreet	281
A funeral elegy upon Mrs. Anne Bradstreet	283
Occasional Meditations:	
For my dear son Simon Bradstreet	291
Meditations divine and moral	292
To my dear children	313
"By night, when others soundly slept"	320
For deliverance from a fever	321
From another sore fit	322
Deliverance from a fit of fainting	323
Meditations when my soul hath been refreshed	324
Upon my son Samuel his going for England	332
For the restoration of my dear husband	334
Upon my daughter Hannah Wiggin	335
On my son's return out of England	335
Upon my husband his going into England	337
In my solitary hours	339
For the letters I received from my husband	341
For my husband's safe arrival	342
"In silent night, when rest I took"	343
"As weary pilgrim, now at rest"	346

ILLUSTRATIONS.

Anne Bradstreet	Frontispiece.
Governor Simon Bradstreet	opp. vii
Chief Justice Joseph Dudley	viii
Chief Justice Paul Dudley	xii
The Bradstreet Residence	xx
Hallway of the Bradstreet House	xxix
Rev. John Cotton	218
John Winthrop	224
John Eliot	228
Sir Philip Sidney	230
William Sallust Du Bartas	234
Extract from the Boston "News Letter"	348

Cotton, Winthrop, and Eliot are inserted here as contemporaneous authors and representative Puritans.

GOVERNOR SIMON BRADSTREET.
Husband of Anne.
From the original painting in the State House, Boston, Mass.

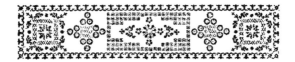

THE POEMS OF MRS. ANNE BRADSTREET.

When it was proposed to me, not long since, to write an introduction to the edition of the poems of Mrs. Anne Bradstreet which "The Duodecimos" were about to issue, many reasons compelled me to decline the task. The request, however, led me to take up once more, after an interval of many years, the poems of "the tenth Muse," as Mrs. Bradstreet was termed on the title-page of the first edition of her verses, and I turned to the elaborate and excellent edition of them published, thirty years ago, by Mr. John Harvard Ellis. After looking them through, I came on the "Elegy upon the truly pious, peerless, and matchless gentlewoman Mrs. Anne Bradstreet," written by my ancestor the Reverend John Norton, of Hingham. I had quite forgotten its existence, and, on reading it, it struck me that there would be something of quaint appropriateness in my writing, at this long interval, in regard to her whose praises he had sung, and that the act would not be without a certain piety toward my ancestor. And, further, I reflected, that as I could trace my descent in one line directly from Governor Thomas Dudley, the father of Mrs. Bradstreet, and as the portraits of her brother, Governor Joseph Dudley, and his wife, looked down on me every day while I sat at breakfast and dinner, she, as my aunt many times removed, might not unjustly have a claim upon me for such token of respect to her memory as had been asked of me. Moved by these pious considerations, I revised my decision.

I am sorry that I cannot speak with admiration of my venerable ancestor Mr. John Norton's verses, but their defects may, in part at least, be excused by his youth at the time when they were written. Mrs. Bradstreet died in 1672, two hundred and twenty-five years ago, and if the Elegy were written at that time (it first appeared in the second edition of her poems in 1678) Mr. Norton was in his twenty-second year, and had graduated at Harvard the year before. His verses are artificial in sentiment, extravagant in expression, and cumbered with pedantry. The Elegy contains, indeed, two tolerably

CHIEF JUSTICE JOSEPH DUDLEY.
Half-brother of Anne (Dudley) Bradstreet.
From the original painting owned by
Professor Charles Eliot Norton, Cambridge, Mass.

good lines, which is not a bad proportion, considering the usual character of such performances, in which a single excellent verse would be surprising; but to my regret I am obliged to acknowledge that these two creditable lines do not belong to the professed author. They are these:

"Like a most servile flatterer he 'll show
Though he write truth and make the subject you."

Now it happens that Francis Beaumont, in the poem addressed by him to the Countess of Rutland, the only daughter of Sir Philip Sidney, had written:

"Although I know whate'er my verses be,
They will like the most servile flattery shew,
If I write truth and make the subject you."

It amused me to find that the young graduate, then engaged in his theological studies, had had recourse to the poems of the playwright, who was not held in good esteem by the devout of those days.

But even Mrs. Bradstreet's repute as a poet, great as it was in her own little circle, hardly stands the test of time, and it is not their poetic merit which will lead any one at the present day to read her verses.

The little that is known of her life has been often told. She and her husband were alike of gentle blood and gentle breeding. She was born in 1612, and married when only sixteen years old to a youth of promise nine years older than herself. Two years later, in 1630, they accompanied her father, Mr. Thomas Dudley, so distinguished in the later history of the Massachusetts colony, on the memorable voyage of Winthrop and his companions in the *Lady Arbella*. Next after Winthrop, Dudley was the foremost man of the emigration, and the young Bradstreet was already one of the "assistants" of the Massachusetts Company, and seems to have been held in respect for his own character, as well as for his relationship to one of the leaders of the party. Of Mrs. Bradstreet during the hard early years of the Massachusetts settlement nothing is recorded, and in her poems she tells us nothing of the events of her life at this time. It is, indeed, a striking fact in regard to her poetry, and a criticism upon it as well, that in it all there is scarcely a reference to New England, and no word from which one might gather that it had been

Introductory

written in the New World at a time so difficult, so interesting, so strange to these new-comers ~~from the Old~~. All her allusions, her figures of speech, her illustrations are drawn from the old worn-out literary stock. No New England bird sings in her pages; ~~it is Philomel, or the lark~~; no New England flower seems to have been dear to her; no incident or aspect of life peculiar to New England is described or even referred to. Nothing can be gathered from her verses in regard to the modes of existence or the social experience of the first emigrants to this "uncouth corner of the world," as Governor Belcher later called it. Of all those things about which we should be curious and interested to hear there is not a word.

It is noteworthy how little of poetic sentiment the New Englanders displayed during the first century of the settlement. There was abundance of religious feeling; abundance of domestic sentiment; a quantity of verse was written; but in the whole mass there is scarcely one line instinct with imagination, and few that show a play of fancy or sustained liveliness of humor. The verses for the most part seem to partake of the rugged character of the land which the English-born settlers were mastering, and if every now and then there be a gleam of humor, as in some of the verses of the eccentric Reverend Mr. Nathaniel Ward, of which an illustration is afforded by the commendatory piece which he prefixed to the first edition of Mrs. Bradstreet's poems — if occasionally, I say, there be a flash of wit or humor, it has no native color, but might as well have had its origin in Old as in New England.

It was not that the colonists were uninstructed people, or that they lacked knowledge of letters; but their minds were occupied mainly with other matters more serious to them than poetry. They were busy in providing for the essential needs of material life, and busier still in saving their souls according to a doctrine which left them little inclination for what seemed to them so trivial an occupation as the making of verse. They were cut off from association with cultivated society, and were remote alike from the current of the intellectual life of the time and from the sources of refinement and of taste. This is strikingly evident even in Mrs. Bradstreet's poems, which, indeed, were the best the first generation of emigrants to New England produced. She had not been deprived of books, for her father was a lover of

CHIEF JUSTICE PAUL DUDLEY.
Son of Joseph, and founder of the Dudleian Lecture
at Harvard College.
From the original painting owned by Dudley R. Child, Esq.,
Boston, Mass.

them,—*Helluo librorum* he is termed in his epitaph,— and he left at his death a small but choice collection of some sixty volumes. She was acquainted with at least three books among the most precious in the whole field of English literature, the "Faërie Queene," the "Arcadia," and North's "Plutarch."[1] But though she refers to Spenser, there is no sign in her verses that she really cared for his poem. Her master in poetry was Du Bartas, in "silver-tongued" Sylvester's translation. She refers often to the delight which she took in his poetry, and to its having been the inspir-

---

[1] Mr. Ellis has given in his Introduction a list of the authors to whom Mrs. Bradstreet refers, or whose works she had probably read. In a note he points out a curious resemblance in one of her verses to words in "Hamlet." It is in the first edition of her poems and was changed in the second, and it occurs near the end of the second of the "Four Ages of Man." The verse stood:

"Ceased [seiz'd] by the gripes of Serjeant Death's arrests,"

which certainly seems to hark back to Hamlet's

" This fell sergeant, death,
Is strict in his arrest," v. ii. 347-8.

It would be of interest to know that Mrs. Bradstreet had read the play. There is, I believe, no evidence that there was a copy of Shakespeare's plays in Massachusetts during the seventeenth century.

ation of her muse. She begins a copy of verses in his honor, with the declaration,—

> "Among the happy wits this age hath shown,
>   Great, dear, sweet Bartas thou art matchless shown,"

and she proceeds to extol him in terms which at last lead her to exclaim,

> "Pardon if I adore, when I admire."

The Reverend Mr. Ward was not far wrong when, in his commendatory verses, he says

> "The Auth'ress was a right *Du Bartas* Girle."

The immense vogue and influence of Du Bartas's poems in France and in England for more than half a century,[1] contrasted with the oblivion into which they have fallen in both countries, affords an illustration not so much of the mutability of taste, as of the fact that circumstances other than its purely poetic merit may sometimes secure for verse an immediate

---

[1] The chief poem of Du Bartas, "The Week, or the Creation of the World," was first published between 1570 and 1580, the exact date is uncertain.

popularity so genuine and so wide-spread as to give a delusive promise of lasting fame. Wordsworth in the essay supplementary to his famous Preface of 1815 asks: "Who is there that can now endure to read 'The Creation' of Du Bartas? Yet all Europe once resounded with his praise; he was caressed by kings; and when his poem was translated into our language, 'The Faery Queen' faded before it." Mr. Lowell, I think, goes too far when, in his essay on Spenser, he declares Wordsworth's statement to be "wholly unfounded." For the moment, and with a large class, the poem of Du Bartas had an acceptance far beyond that of Spenser, but Mr. Lowell is right when he adds that "the vitality of a poem is to be measured by the kind as well as the amount of influence it exerts." Spenser himself in the "L'Envoy" to his translation of Du Bellay's "Ruines of Rome" speaks of the "heavenly sense" of Du Bartas, and uniting him with Du Bellay, exclaims:

"Live happie Spirits, th' honour of your name,
And fill the world with never dying fame!"

Du Bartas's "Creation" retained its popularity well through the whole Puritan period. In his dedication of the "Spanish Friar," in 1681, Dryden says: "I remember when I was a boy I thought inimitable Spenser a mean poet in comparison of Sylvester's Du Bartas, and was rapt into an extasy when I read these lines:

> 'Now when the winter's keener breath began
>   To crystallize the Baltic oceān,
>   To glaze the lakes, to bridle up the floods,
>   And perriwig with snow the baldpate woods.'

I am much deceived if this be not abominable fustian." And in his "Art of Poetry," published in 1683, he again refers to the favorite of his callow days, and, scoffing at him with a lively quip, says:

> "Thus in times past Dubartas vainly writ,
>   Allaying sacred truth with trifling wit;
>   Impertinently, and without delight,
>   Describ'd the Israelites' triumphant flight,
>   And following Moses o'er the sandy plain,
>   Perish'd with Pharaoh in the Arabian main."

And again in the same poem, warning against bombast and fustian, he cites afresh the verses which had once charmed him, and bids the poets

Introductory xvii

"Not with Du Bartas 'bridle up the floods,
And perriwig with wool the baldpate woods.'"[1]

Even in our own century Du Bartas has not been without admirers who have tried to restore credit to his work, and, surprising as it may seem, chief among them is Goethe. He rebukes the French for their contempt and neglect of "The Creation," and declares that it possesses genuine elements of poetry, though strangely mingled. The author, he says, deals with weighty and important themes which afford him opportunity to display a naïve view of the world, and to exhibit entertainingly, in description, narrative, and didactic discourse, an immense variety of knowledge.[2]

In this characterization of the poem Goethe undoubtedly accounts in part for its popularity in the early seventeenth century. Even Sainte-Beuve, who contests Goethe's judgment, admits that Du Bartas had a certain Bœotian fertility, and that fine fragments may be detached from the mass of "his dispropor-

---

[1] These lines which ran in Dryden's head are to be found in the Fourth Book of the First Day of the Second Week of the Creation. His first citation of them is wrong in substituting "snow" for "wool."

[2] See the Notes appended to "Rameau's Neffe."

tioned Babel." The vogue of "The Creation" was, indeed, due rather to its encyclopedic character, and to its poetic faults, than to its poetic merit. It was a compilation of miscellaneous learning, and while it did not lack spirit in versification and abounded in ingenious imagery, its very extravagance of diction and excess of conceits suited the general taste. But there was a still deeper reason for its wide acceptance. "The Creation" combined piety with entertainment; it was the work of a grave disciple of Calvin, who invoked the Christian muse in opposition to the pagan mistresses of Ronsard and his followers. It was a poem for men who cared more for purity of doctrine than for purity of poetry, for men more interested in the Bible than in profane literature; it was the poem of a party in religion. The version by Sylvester, which was published in 1605, preserved essentially the character of the original, and there is no reason for wonder that a Puritan girl, born in 1612, when Milton was four years old, a girl bred piously and strictly, yet inspired with some faint poetic instinct, should have found delight as well as instruction in Du Bartas's verse, and should have taken him for her master in the divine art.

I have said that Mrs. Bradstreet apparently did not care for Spenser's poetry. She seems to have cared more, but with great reservations, for the "Arcadia." In her elegy upon Sir Philip Sidney, in the midst of her eulogy of him, she says of the "Arcadia":

"I praise thee not for this, it is unfit,
    This was thy shame, O miracle of wit." [1]

And yet, as if repenting of this condemnation, she adds, and the lines are among her most vigorous,—

"But he 's a beetlehead that can't descry
    A world of wealth within that rubbish lie;
    And doth his name, his work, his honour wrong,
    The brave refiner of our English tongue,
    That sees not learning, valor and morality,
    Justice, friendship and kind hospitality,
    Yea and divinity within his book."

She made some little use of North's "Plutarch" in her poem on "The Four Monarchies." But this poem is mainly a mere dry abridgment of Raleigh's

---

[1] These verses are omitted in the second edition of her Poems, and in their place is the line:

"His wiser days condemn'd his witty works."

"History of the World," which indeed I should have added to the list of noble books at her command. The poem is entirely tedious, being little more than a rhymed summary of events, with no spirit of exaltation in recounting the fates of great nations, and no touch of animation in the narrative of the heroic deeds of individuals. It is a long, conscientious, laborious work, which nobody perhaps will ever again read through. I have looked in vain up and down its pages to find a verse which has the genuine tower-stamp of poetry; but I have not found one. Nor have I found in any of her poems the grace and charm of spontaneous lyrical utterance. Every now and then a single verse shows a true, if slight, capacity for poetic expression; as, for instance, in the poem on "The Four Elements":

"My pearls that dangle at thy darling's ears"

is a verse not without melodious flow; and in her poem on "The Four Ages of Man" there is one in which a familiar epithet of Gray's is anticipated:

"But waking, glad to hear the cock's shrill voice";

THE BRADSTREET RESIDENCE AT NORTH ANDOVER, MASS.

Built by Governor Bradstreet to replace the dwelling which was burned in 1666 (see page 343), and said to have been the home of Anne Bradstreet

and there are four verses in her poem on "The Four Seasons" which have more than once been cited as examples of her poetry at its best:

> "The fearful bird his little nest now builds
> In trees and walls, in cities and in fields;
> The outside strong, the inside warm and neat,
> A natural artificer complete."

But if these be the best, what can be said about the large remainder?

Her chief work is a series of four compositions, on "The Four Elements," "The Four Humours in Man's Constitution," "The Four Ages of Man," "The Four Seasons of the Year." They are all of one design: each Element, each Humour, each Age, each Season, is represented as discoursing of itself, setting forth its own good and evil qualities. The scheme is prosaic, but it admits of a great variety of theme and of the display of an unusual amount of knowledge on many subjects. The imitation of Du Bartas is manifest, but it serves rather to enhance the merit of his verses than to secure excellence for those of his admirer. Mrs. Bradstreet has nothing of the energy and abundance of his vein, nothing of the picturesqueness of his broad stream of verse, and her acquisitions—large, even remarkable for a woman in her time and circumstances—were inconsiderable in comparison with his vast if superficial learning. The best of her longer poems is called "Contemplations." It is a series of simple religious reflections on the beauty of nature, the goodness of God, the transiency of man's life and of earthly things. Several of its stanzas of seven lines have grace and ease, and occasional metrical felicity; and had all her work possessed like excellence it might still be read with pleasure. But while the greater part of her poems show good sense and good feeling, and, at times, something of ingenuity and skill, they are devoid of inspiration, and even of the lower enthusiasm of the understanding. They are generally bald and prosaic, and their reader readily accepts her assertion concerning them:

"And for the same I hours not few did spend,
 And weary lines, though lank, I many penn'd."

Mrs. Bradstreet's modest consciousness of the slenderness of her poetic outfit is, indeed, such as to show that she was a better judge of her verses than

Introductory xxiii

her too partial friends. The first edition of her poems was published in London in 1650, by her brother-in-law, the Rev. John Woodbridge of Andover, then on a visit to the old country. He says in his preface to the volume, "I fear the displeasure of no person in the publishing of these poems but the Author's, without whose knowledge, and contrary to her expectation, I have presumed to bring to public view what she resolved should never in such a manner see the sun." In a little piece entitled "The Author to her Book," written apparently with a view to a second edition of it, and which has more fancy in it than any other which she ever wrote, Mrs. Bradstreet expresses with a pretty simplicity her feeling at seeing in print "the ill-formed offspring of her feeble brain." Let the reader turn to this little poem, and he will gain a very kindly feeling for the gentle lady who wrote it, while its last verses will interest him as a native specimen of the "Envoy" with which the poets of the day were wont to send forth their work.

"In better dress to trim thee was my mind,
But naught save home-spun cloth i' th' house I find;

In this array, 'mongst vulgar mayst thou roam,
In critics' hands beware thou dost not come;
And take thy way where yet thou art not known;
If for thy Father askt, say, thou hadst none,
And for thy Mother, she, alas, is poor,
Which caus'd her thus to send thee out of door." [1]

Some of Mrs. Bradstreet's occasional poems possess a charm of natural and simple feeling which still touches the heart. They are the expressions of her domestic sentiment, addressed to her husband or to her children; or hymns in which she utters the devout aspirations and desires of her soul. In a little paper of religious experiences, which she prepared late in life as a legacy to her children, there is a passage which makes one wish that she had put more of her own personal experience into her verse. She says: "About sixteen the Lord laid his hand sore upon me and smote me with the smallpox. When I was in my affliction I besought the Lord, and con-

---

[1] These verses recall those of Spenser "To his Book," prefixed to the "Shepherds' Calendar":

"But if that any ask thy name
Say thou wert base begot with blame."

fessed my pride and vanity, and He was entreated of me and again restored me; but I rendered not to Him according to the benefit received. After a short time I changed my condition, and was married, and came into this country, where I found a new world and new manners, at which my heart rose. But after I was convinced it was the way of God I submitted to it and joined the church at Boston." Would that she had told us of the trials of that time, and why it was that her heart rose against the new world and the new manners to which she had come!

Besides this reference to this early hard experience, there is nothing in Mrs. Bradstreet's papers to indicate that she suffered, as so many of the women of her time and later suffered, from the black doctrine which made their lives dark with its shadow. Her religious meditations have remarkable sweetness and simplicity, and express a confidence in the mercies of God which it was seldom given to the tender-hearted in those days to attain. Something of this spirit no doubt was due to the native serenity and tranquillity of her disposition. She even fronted with calmness the dreadful peril of atheism which dismayed so many souls, and she says very simply with regard to it, that, "many times hath Satan troubled me concerning the verity of the Scriptures, many times by atheism, how I could know whether there was a God. I never saw any miracles to confirm me, and those which I read of, how did I know but they were feigned. That there is a God my reason would soon tell me by the wondrous works that I see, the vast frame of the heaven and the earth, the order of all things, night and day, summer and winter, spring and autumn, the daily providing for this great household upon the earth, the preserving and directing of all to its proper end. The consideration of these things would with amazement constantly resolve me that there is an eternal being." This is unusual thinking and unusual writing for a New England woman of the first generation.

Her life must have been occupied mainly with household cares, for she became the mother of eight children, all of whom lived to grow up. But of the special incidents of that life there are few indications either in her poems or in the remains of her prose. One event affected her greatly. In the year 1666, in July, not quite two months before the Great Fire of London, her own house in the pleasant township

Introductory

of Andover, which had been her home for some twenty years, was burned. There are touches of natural feeling in the verses which she wrote on the occasion, and one sympathizes with her when, looking at the ruins, she reflects:

> "Here stood that trunk, and there that chest,
> There all that store I counted best;
> My pleasant things in ashes lie,
> And them behold no more shall I.
> Under thy roof no guest shall sit,
> Nor at thy table eat a bit.
>
> "No pleasant tale shall e'er be told,
> Nor things recounted done of old,
> No candle e'er shall shine in thee,
> Nor bridgroom's voice e'er heard shall be;
> In silence ever shalt thou lie,
> Adieu, adieu, all 's vanity."

Among the things which perished in the burning, and which she perhaps regretted more than others of more worth, was the conclusion of her poem on the Four Monarchies. It remained unfinished at her death.

But little as we know of her daily occupations and interests, and difficult as it is to follow even in fancy the daily life of a housewife in New England in that early time, there is enough in the pieces addressed to her husband and to her children to indicate that in her home was much affection and much happiness. Her husband, according to such report as has come down to us regarding him, was an intelligent and well-intentioned man, a conscientious Puritan, trustworthy in affairs, and of a kindly disposition. He does not seem to have been distinguished by superior talents, but he had a character which secured the respect and confidence of his associates. His wife writes to him in terms such as she could not have used if she had not found in him all that was needed to make her content with life. He long survived her, living to be ninety-four years old, thus acquiring and deserving the appellation of the Nestor of the Colony. She begins a poem, "To my dear and loving Husband," with the words:

> "If ever two were one, then surely we,
>   If ever man were loved by wife, then thee;
>   If ever wife was happy in a man,
>   Compare with me, ye women, if ye can."

Other poems addressed to him are less simple in expression than this, and in them she indulges in the

HALLWAY OF THE BRADSTREET HOUSE
AT NORTH ANDOVER, MASS.
From a photograph, 1896.

conceits which were favored by poets of her time. Perhaps the most amusing is one derived from her favorite Du Bartas, who, in his account of the fishes in the Fifth Day of the Creation, tells how the mullet was distinguished above all other creatures for its fidelity to its mate. So Mrs. Bradstreet, writing to her husband, says:

"Return my dear, my joy, my only love,
Unto thy hind, thy mullet, and thy dove,
Who neither joys in pasture, house, nor streams;
The substance gone, O me, these are but dreams.
Together at one tree, oh let us browse,
And like two turtles roost within one house,
And like the mullets in one river glide —
Let's still remain but one, till death divide.

Thy loving love, and dearest dear,
At home, abroad and everywhere."

But perhaps of all her domestic poems there is none which has a truer accent of emotion than one written "On my Son's return out of England, July 17, 1661." The son had been away for more than four years, and she begins her verses with,

> "All praise to Him who now hath turned
> My fears to joys, my sighs to song,
> My tears to smiles, my sad to glad:
> He 's come for whom I waited long."

And again in the next year, when her husband returned from a visit to England, whither he, with the Rev. John Norton, the elder, had been sent on an important mission as agents of the Colony, she breaks out into praises to the Lord with,

> "What shall I render to Thy Name,
> Or how Thy praises speak?
> My thanks how shall I testify?
> O Lord, thou know'st I 'm weak."

Such utterances are witnesses alike of the depth of her piety and the strength of her affections.

I said just now that it was difficult for us to reconstruct in imagination the days of the New England woman of the first generation transplanted from the Old World. Our lives are too remote from theirs in all external conditions to enable us to picture save in outline the interests and the occupations with which they were most concerned. But it is not difficult to form the image of a character like Mrs. Bradstreet's as it is shown in her own writing, under the conditions of life which we know must have existed for her. It is the image of a sweet, devout, serene, and affectionate nature, of a woman faithfully discharging the multiplicity of duties which fell upon the mother of many children in those days when little help from outside could be had; when the mother must provide for all their wants with scanty means of supply, and must watch over their health with the consciousness that little help from without was to be had in case of even serious need. I fancy her occupying herself in the intervals of household cares with the books which her own small library and her father's afforded, and writing, with pains and modest satisfaction, the verses which were so highly esteemed at the time, but which for us have so little intrinsic interest. She cherished in herself and in her children the things of the mind and of the spirit; and if such memory as her verses have secured for her depend rather on the rare circumstance of a woman's writing them at the time when she did, and in the place where she lived, than upon their poetic worth, it is a memory honorable to her, and it happily preserves the name of a good woman, among whose descendants has been more than one poet whose verses reflect lustre on her own.[1]

CHARLES ELIOT NORTON.

JANUARY, 1897.

---

[1] Through one of her children she is the ancestress of Richard Henry Dana; through another, of Oliver Wendell Holmes.

EDITOR'S NOTE.

The FIRST EDITION of Mrs. Bradstreet's poems was printed in London in 1650. There had been a press at Cambridge, Massachusetts, since 1638, but there is no reason to suppose that this book was offered to it for printing, for the press was constantly occupied with church, state, and educational documents of importance, and had no leisure for work which was not of stern necessity.

It would seem that the Rev. John Woodbridge, who had come to New England in 1634, and had married Mrs. Bradstreet's younger sister Mercy, was much impressed by his sister-in-law's "gracious demeanor, eminent parts, pious conversation, and courteous disposition," and, upon his return for a visit to the mother country in 1647, took with him a number of her poems in manuscript, and had them printed in London without the consent of the author. To justify himself in his course, he secured a number of commendatory epistles in verse from friends and ad-

THE TENTH MUSE

Lately sprung up in AMERICA.

OR

Severall Poems, compiled with great variety of VVit and Learning, full of delight.

Wherein especially is contained a compleat discourse and description of

The Four
- Elements,
- Constitutions,
- Ages of Man,
- Seasons of the Year.

Together with an Exact Epitomie of the Four Monarchies, viz.

The
- Assyrian,
- Persian,
- Grecian,
- Roman.

Also a Dialogue between Old *England* and New, concerning the late troubles.

With divers other pleasant and serious Poems.

By a Gentlewoman in those parts.

Printed at London for *Stephen Bowtell* at the signe of the Bible in Popes Head-Alley. 1650.

TITLE-PAGE OF THE FIRST EDITION.

SIZE OF THE ORIGINAL.

mirers of the author, and inserted them at the beginning of the volume directly after his own quaint preface in which he sought to appease the expected resentment of Mrs. Bradstreet.

"The Tenth Muse" could not have been a woman if when she received a copy of the book she did not seize upon it, in spite of her protestations, with a fluttering, pleased excitement. But a perusal of her writings in type revealed to her mortified gaze the extent of her own shortcomings and the inevitable blunders of the printer. Mrs. Bradstreet was the first — but not the last — American author whose "blushing was not small" at sight of her first book; and she later (p. 266 of this edition) recorded with some asperity her feelings against those "friends less wise than true" who were responsible for the publication of her "ragged lines," and against the printer who instead of "lessening her errors" added fresh faults of his own.

She undertook a revision of this edition, but with the birth of her eighth child, the death of her father, the frequent absence of her husband upon public employment, and her family cares, her literary occupations were interrupted; and when in July, 1666, the house in which she lived at Andover was burned to the ground, and her papers "fell a prey to the raging fire," she seems to have abandoned all idea of further effort in that direction.

Mrs. Bradstreet died in 1672. Six years later the SECOND EDITION of her Poems was printed. At the end of the book was placed additional matter, with this heading: "*Several other poems made by the author upon divers occasions were found among her papers after her death, which she never meant should come to public view; amongst which these following, at the desire of some friends that knew her well, are here inserted.*" It is surmised that this edition was prepared for the press by the Rev. John Norton, of Hingham, who appended a "Funeral Elegy" upon the author.

John Foster was the printer of this book. He was graduated from Harvard College in 1667, was authorized to set up a press at Boston about 1676, and died in 1681 aged thirty-three years. Although Mr. Foster appears to have been much respected, he was responsible for what may be called "a deal of indifferent printing." Just what part he took in the actual labor of book-making is not known; it is charitable to suppose that he was not bred to the art, and employed unskilful and careless workmen.

In setting the types for this Second Edition a certain measure, or width of page, was chosen. This was roomy enough for the majority of lines in the book; but an occasional long-syllable verse was met, and the compositor seems to have tried hard to make each one fit the measure, and not to allow a portion of it to turn over to make another line. Thus we

SEVERAL
POEMS

Compiled with great variety of Wit and
Learning, full of Delight.
Wherein especially is contained a compleat
Discourse, and Description of

The Four { ELEMENTS
CONSTITUTIONS,
AGES of Man,
SEASONS of the Year.

Together with an exact Epitome of
the three first *Monarchyes*

Viz. The { ASSYRIAN,
PERSIAN,
GRECIAN.

And beginning of the Romane Common-wealth
to the end of their last *King*:

With diverse other pleasant & serious *Poems*;

By a Gentlewoman in *New-England*.

*The second Edition, Corrected by the Author,
and enlarged by an Addition of several other
Poems found amongst her Papers
after her Death.*

Boston, Printed by *John Foster*, 1678.

TITLE-PAGE OF THE SECOND EDITION.

*C   SIZE OF THE ORIGINAL.

frequently find words in such a case huddled together with very little space between them,— sometimes none at all; and when that did not avail words were abbreviated, with or without an apostrophe, capitals were reduced to small letters, long "and" was replaced by short "&," punctuation marks were omitted, and other devices applied to accomplish the purpose. This heroic treatment was the common resort of compositors in the early stages of typography, and had not fallen entirely into disuse in the seventeenth century, as a few examples from the Second Edition of Mrs. Bradstreet's poems will show; the first line of each couplet gives normal typography, the second as Mr. Foster printed it.

He peer'd, and por'd, and glar'd, and said, for wore,
He peer'd, and por'd, & glar'd, & said for wore,

(page v)

Earth, thou hast not moe countrys, vales, and mounds
Earth thou hast not moe countrys vales & mounds

(page 16)

Laughter (though thou say malice) flows from hence,
Laughter (thô thou say malice) flows from hence,

(page 36)

Now up, now down, now chief, and then brought under.
Now up now down now chief, & then broght under,

(page 184)

And ſmote thoſe feet, thoſe legs, thoſe arms, and thighs,
and ſmote thoſe feet thoſe legs, thoſe arms & thighs

(page 185)

Foster's fonts of type were not large. At certain points in each signature he ran out of various sorts: hence we frequently find VV for W, although the latter was in abundance in pages immediately preceding; italic instead of roman marks of punctuation, as, *:* *;* *?*; and upturned commas in place of apostrophes, as, "o'er," "call'd,"—not in isolated cases, but in patches of some extent which leave no doubt of the reason for their being. In one place capital I is replaced by lower case i, and at another point by lower case l, in such a way as to appear intentional because he had not enough of the capitals. Although the apostrophe was used on every page to show elision of letters, as, "ne'er," "you'll," "auth'ress," there is no instance of its use to indicate the possessive case, as, "Authors wit," "womans wrath," "Chaucers boots." Capitals were used very profusely, and without method; on some pages few appear, on others nearly all nouns are headed with them. A specimen line is this:

' His Suit of Crimson and his scarfe of green"

(page 44)

SEVERAL
POEMS

Compiled with great Variety of WIT and LEARNING, full of DELIGHT;

Wherein especially is contained, a compleat Discourse and Description of

The Four { ELEMENTS, CONSTITUTIONS, AGES of MAN, SEASONS of the Year.

Together with an exact EPITOME of the three first MONARCHIES, viz. the

ASSYRIAN, PERSIAN, GRECIAN, and ROMAN Common WEALTH, from its beginging, to the End of their last KING.

With divers other pleasant and serious POEMS.

By a GENTLEWOMAN in *New-England.*

The THIRD EDITION, *corrected by the Author, and enlarged by an Addition of several other* POEMS *found amongst her Papers after her Death.*

Re-printed from the second Edition, in the Year
.M.DCC.LVIII.

TITLE-PAGE OF THE THIRD EDITION.

SIZE OF THE ORIGINAL.

The THIRD EDITION was printed also at Boston, in 1758. The names of the publisher and printer are not known. It was a reprint from the second edition, though by a simple change in the title-page the impression is given that Mrs. Bradstreet was responsible for the numerous corrections of spelling and capitalization and other improvements found therein. Of course the Third Edition was *not* "corrected by the Author." The types in this edition were more accurately composed than in either of those preceding.

The FOURTH EDITION was printed at Cambridge in 1867 for Abram E. Cutter, of Charlestown, under the supervision of John Harvard Ellis, of Boston. Mr. Ellis was most painstaking in his labors; his researches were original and extensive, his references authoritative, and his notes helpful. He included a quantity of material not in the other editions,—undoubtedly the remainder of that which was found among Mrs. Bradstreet's papers after her death, and which "she never meant should come to public view."

There is ground for but one difference of opinion with Mr. Ellis. He reprinted the writings of Mrs. Bradstreet after the Second Edition, retaining carefully all its wretched spelling, confusing punctuation,

THE WORKS OF

Anne Bradstreet

IN PROSE AND VERSE

EDITED BY

JOHN HARVARD ELLIS

Charlestown

ABRAM E. CUTTER

1867

TITLE-PAGE OF THE FOURTH EDITION.

REDUCED SIZE.

unmethodical capitalization, and even its typographical errors! The Second Edition was replete with unintentional errors, mere results of carelessness on the part of the printer; as, "Is is possible?" for "Is it possible?" "Snrdanapal" for "Sardanapal," "tortnr'd" for "tortur'd," "Perslan" for "Persian," "feblee" for "feeble," "strenght" for "strength," and so *ad finem*. Mr. Ellis scrupulously reproduced these plain misprints. He even inserted a roman letter in an italic word on the same authority, thus, "*New-England*"; and as to the indications of a small font heretofore described, such as the use of VV for W, italic punctuation marks for roman, upturned commas for apostrophes, lower case i and l for capital I,— in all these, *mirabile dictu*, the Fourth Edition followed its leader.

This seems unreasonable, and The Duodecimos, in preparing the FIFTH EDITION, determined to print these writings of the first American poet as though they had never been printed before. They here offer a volume in which the orthography, especially of proper names, has been carefully modernized, in which evident printers' errors have been corrected, and a few trifling alterations made to avoid perpetuating instances of unnecessarily bad grammar,— defects which can add no value to a new edition, but which obscure such meaning as the lines may contain. As a rule, elided letters have been supplied in words like "hind'ring," "heav'n," "to 't," and "th'," in accordance with general modern usage. The reader will understand that the meter requires the slurred pronunciation formerly indicated by the apostrophe.

For permission to use the portraits and other illustrations included in this volume the especial thanks of The Duodecimos are extended to the Commonwealth of Massachusetts; the American Antiquarian Society, Worcester, Mass., which also kindly placed in the hands of the Publication Committee for comparison its perfect copy of the rare Second Edition of Mrs. Bradstreet's poems; the Lenox Library, New York; Prof. Charles Eliot Norton, Cambridge, Mass.; Mr. Dudley R. Child, Boston, Mass.; and Mr. Hollis R. Bailey, of Boston. To very many other gentlemen grateful acknowledgment of their interest is hereby expressed.

FRANK E. HOPKINS,
For the Publication Committee.

ERRATA.

Page 212, line 10, for "coats" read "cotes."
Page 287, line 7, for *"fileant"* read *"sileant."*

[The poems following to page 287 inclusive are reprinted from the Second Edition, 1678, of the writings of Mrs. Bradstreet. The facsimile of the title-page of that edition, printed on page xxxvii of this volume, is repeated in this connection as part of the text.]

SEVERAL
POEMS

Compiled with great variety of Wit and Learning, full of Delight.

Wherein especially is contained a compleat Discourse, and Description of

The Four { ELEMENTS, CONSTITUTIONS, AGES of Man, SEASONS of the Year.

Together with an exact Epitome of the three first *Monarchyes*

Viz. The { ASSYRIAN, PERSIAN, GRECIAN.

And beginning of the Romane Common-wealth to the end of their last King:

With diverse other pleasant & serious *Poems*;

By a Gentlewoman in *New-England*.

The second Edition, Corrected by the Author, and enlarged by an Addition of several other Poems found amongst her Papers after her Death.

Boston, Printed by *John Foster*, 1678.

FACSIMILE OF THE TITLE-PAGE OF THE SECOND EDITION.

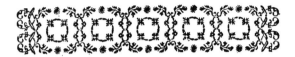

Kind Reader:

Had I opportunity but to borrow some of the Author's wit, 't is possible I might so trim this curious work with such quaint expressions as that the preface might bespeak thy further perusal; but I fear 't will be a shame for a man that can speak so little to be seen in the title-page of this woman's book, lest by comparing the one with the other the reader should pass his sentence that it is the gift of women not only to speak most but to speak best. I shall leave therefore to commend that which with any ingenuous reader will too much commend the Author, unless men turn more peevish than women, to envy the excellency of the inferior sex. I doubt not but the reader will quickly find more than I can say, and the worst effect of his reading will be unbelief, which will make him question whether it be a woman's work, and ask, Is it possible? If any do, take this as an answer from him that dares avow it: It is the work of a woman, honored and esteemed where she lives for her gracious demeanor, her eminent parts, her pious conversation, her courteous disposition, her exact diligence in her place, and discreet managing of her family occasions; and more than so, these poems are the fruit of but some few hours curtailed from her sleep and other refreshments. I dare add little lest I keep thee too long. If thou wilt not believe the worth of these things in their kind when a man says it, yet believe it from a woman when thou seest it. This only I shall annex: I fear the displeasure of no person in the publishing of these poems but the Author, without whose knowledge, and contrary to her expectation, I have presumed to bring to public view what she resolved in such a manner should never see the sun; but I found that divers had gotten some scattered papers, affected them well, were likely to have sent forth broken pieces to the Author's prejudice, which I thought to prevent, as well as to pleasure those that earnestly desired the view of the whole.

Mercury showed Apollo Bartas' book,
Minerva this, and wished him well to look
And tell uprightly which did which excel.
He viewed and viewed, and vowed he could not tell.
They bid him hemisphere his moldy nose
With his cracked leering glasses, for it would pose
The best brains he had in his old pudding-pan,
Sex weighed, which best — the woman, or the man?
He peered, and pored, and glared, and said, forwore,
" I 'm e'en as wise now as I was before."
They both 'gan laugh, and said it was no mar'l,
The Authoress was a right Du Bartas girl.
" Good sooth!" quoth the old Don, " tell ye me so?
I muse whither at length these girls will go.
It half revives my chill frost-bitten blood
To see a woman once do aught that 's good;
And chode by Chaucer's boots and Homer's furs,
Let men look to it lest women wear the spurs.
                                             N. WARD.

TO MY DEAR SISTER,
THE AUTHOR OF THESE POEMS.

Though most that know me dare, I think, affirm
I ne'er was born to do a poet harm,
Yet when I read your pleasant witty strains
It wrought so strongly on my addle brains
That though my verse be not so finely spun,
And so like yours cannot so neatly run,
Yet am I willing, with upright intent,
To show my love without a compliment.
There needs no painting to that comely face
That in its native beauty hath such grace.
What I, poor silly I, prefix, therefore,
Can but do this, make yours admired the more;
And if but only this, I do attain
Content that my disgrace may be your gain.

If women I with women may compare,
Your works are solid, others' weak as air.
Some books of women I have heard of late,
Perusèd some, so witless, intricate,
So void of sense and truth as if to err
Were only wished, acting above their sphere;

To My Dear Sister

And all to get what, silly souls, they lack—
Esteem to be the wisest of the pack.
Though, for your sake, to some this be permitted
To print, yet wish I many better witted;
Their vanity makes this to be inquired,
If women are with wit and sense inspired.
Yet when your works shall come to public view,
'T will be affirmed, 't will be confirmed, by you.
And I, when seriously I had revolved
What you had done, I presently resolved
Theirs was the persons', not the sex's, failing,
And therefore did bespeak a modest vailing.
You have acutely, in Eliza's ditty,
Acquitted women, else I might with pity
Have wished them all to women's works to look,
And never more to meddle with their book.
What you have done the sun shall witness bear
That for a woman's work 't is very rare;
And if the Nine vouchsafe the Tenth a place,
I think they rightly may yield you that grace.

But lest I should exceed, and too much love
Should too too much endeared affection move
To superadd in praises, I shall cease,
Lest while I please myself I should displease
The longing reader, who may chance complain,
And so requite my love with deep disdain,
That I, your silly servant, stand in the porch,
Lighting your sunlight with my blinking torch;

Hindering his mind's content, his sweet repose,
Which your delightful poems do disclose
When once the casket's opened.   Yet to you
Let this be added, then I 'll bid adieu:
If you shall think it will be to your shame
To be in print, then I must bear the blame.
If it be a fault, 't is mine; 't is shame that might
Deny so fair an infant of its right
To look abroad.   I know your modest mind:
How you will blush, complain 't is too unkind
To force a woman's birth, provoke her pain,
Expose her labors to the world's disdain.
I know you 'll say you do defy that mint
That stamped you thus to be a fool in print.
'T is true, it doth not now so neatly stand
As if 't were polished with your own sweet hand;
'T is not so richly decked, so trimly attired;
Yet it is such as justly is admired.
If it be folly, 't is of both or neither:
Both you and I, we 'll both be fools together;
And he that says 't is foolish, if my word
May sway, by my consent shall make the third.
I dare outface the world's disdain for both
If you alone profess you are not wroth.
Yet, if you are, a woman's wrath is little
When thousands else admire you in each tittle.

           I. W.

UPON THE AUTHOR.
BY A KNOWN FRIEND.

Now I believe tradition, which doth call
The Muses, Virtues, Graces, females all;
Only they are not nine, eleven, nor three —
Our Authoress proves them but one unity.
Mankind, take up some blushes on the score;
Monopolize perfectiō no more;
In your own arts confess yourselves outdone:
The moon hath totally eclipsed the sun —
Not with her sable mantle muffling him,
But her bright silver makes his gold look dim:
Just as his beams force our pale lamps to wink,
And earthly fires within their ashes shrink.

B. W.

I cannot wonder at Apollo now,
That he with female laurel crowned his brow:
That made him witty! Had I leave to choose,
My verse should be a page unto your Muse.

C. B.

IN PRAISE OF THE AUTHOR, MISTRESS ANNE BRADSTREET, VIRTUE'S TRUE AND LIVELY PATTERN, WIFE OF THE WORSHIPFUL SIMON BRADSTREET, ESQ.; AT PRESENT RESIDING IN THE OCCIDENTAL PARTS OF THE WORLD IN AMERICA, ALIAS NOVA-ANGLIA.

What golden splendent star is this so bright,
One thousand miles twice told, both day and night,
From the orient first sprung, now from the west
That shines, swift-wingéd Phœbus and the rest
Of all Jove's fiery flames surmounting far
As doth each planet every falling star?—
By whose divine and lucid light most clear
Nature's dark secret mysteries appear,
Heaven's, earth's, admired wonders, noble acts
Of kings and princes, most heroic facts,
And whate'er else in darkness seemed to die;
Revives all things so obvious now to the eye
That he who these its glittering rays views o'er
Shall see what was done in all the world before.

N. H.

UPON THE AUTHOR.

'T were extreme folly should I dare attempt
To praise this Author's worth with compliment.
None but herself must dare commend her parts
Whose sublime brain's the synopsis of arts.
Nature and skill here both in one agree
To frame this masterpiece of poetry.
False Fame, belie their sex no more. It can
Surpass, or parallel, the best of man.

C. B.

ANOTHER TO MRS. ANNE BRADSTREET, AUTHOR OF THIS POEM.

I 've read your poem, lady, and admire
Your sex to such a pitch should e'er aspire.
Go on to write; continue to relate
New histories of monarchy and state;
And what the Romans to their poets gave
Be sure such honor and esteem you 'll have.

H. S.

AN ANAGRAM.

Anna Bradestreate.   Deer neat *An Bartas.*

So Bartas-like thy fine spun poems been,
That Bartas' name will prove an epicene.

ANOTHER.

Anne Bradstreate.   Artes bred neat *An.*

UPON MRS. ANNE BRADSTREET, HER POEMS, ETC.

Madam, twice through the Muses' grove I walked;
Under your blissful bowers I shrouding there,
It seemed with nymphs of Helicon I talked,
For there those sweet-lipped Sisters sporting were.
Apollo with his sacred lute sat by.
On high they made their heavenly sonnets fly,
Posies around they strewed of sweetest poesy.

Twice have I drunk the nectar of your lines,
Which high sublimed my mean-born phantasy.
Flushed with these streams of your Maronean wines,
Above myself rapt to an ecstasy,
Methought I was upon Mount Hybla's top,
There where I might those fragrant flowers lop
Whence do sweet odors flow and honey-spangles drop.

To Venus' shrine no altars raiséd are,
Nor venomed shafts from painted quiver fly;
Nor wanton doves of Aphrodite's car
Are fluttering there, nor here forlornly lie
Lorn paramours; nor chatting birds tell news
How sage Apollo Daphne hot pursues,
Or stately Jove himself is wont to haunt the stews.

Nor barking satyrs breathe, nor dreary clouds,
Exhaled from Styx, their dismal drops distil

Upon Mrs. Anne Bradstreet 13

Within these fairy flowery fields, nor shrouds
The screeching night-raven with his shady quill;
But lyric strings here Orpheus nimbly hits,
Arion on his saddled dolphin sits,
Chanting as every humor, age, and season fits.

Here silver swans with nightingales set spells
Which sweetly charm the traveler, and raise
Earth's earthéd monarchs from their hidden cells,
And to appearance summon lapséd days.
There heavenly air becalms the swelling frays,
And fury fell of elements allays
By paying every one due tribute of his praise.

This seemed the site of all those verdant vales
And purléd springs whereat the nymphs do play,
With lofty hills where poets rear their tales
To heavenly vaults which heavenly sound repay
By echo's sweet rebound.   Here ladies kiss,
Circling, nor songs, nor dance's circle miss;
But whilst those sirens sung, I sunk in sea of bliss.

Thus weltering in delight, my virgin mind
Admits a rape; truth still lies undescried.
It 's singular that plural seemed, I find;
'T was Fancy's glass alone that multiplied;
Nature with Art so closely did combine,
I thought I saw the Muses' treble trine,
Which proved your lonely Muse superior to the Nine.

Your only hand those poesies did compose;
Your head the source whence all those springs did flow;
Your voice whence change's sweetest notes arose,
Your feet that kept the dance alone, I trow.
Then vail your bonnets, poetasters all,
Strike lower amain, and at these humbly fall,
And deem yourselves advanced to be her pedestal.

Should all with lowly congees laurels bring,
Waste Flora's magazine to find a wreath,
Or Peneus' banks, 't were too mean offering.
Your Muse a fairer garland doth bequeath
To guard your fairer front: here 't is your name
Shall stand emmarbled; this your little frame
Shall great colossus be, to your eternal fame.

I 'll please myself, though I myself disgrace.
What errors here be found are in Errata's place.

J. Rogers.

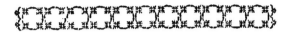

TO HER MOST HONORED FATHER
THOMAS DUDLEY, ESQ.,
THESE HUMBLY PRESENTED.

Dear Sir, of late delighted with the sight ⎰ T.D. *On*
Of your four Sisters clothed in black and ⎱ *the Four*
    white,   *Parts of*
Of fairer dames the sun ne'er saw the face, *the World.*
Though made a pedestal for Adam's race.
Their worth so shines in these rich lines you show,
Their parallels to find I scarcely know.
To climb their climes I have nor strength nor skill;
To mount so high requires an eagle's quill.
Yet view thereof did cause my thoughts to soar —
My lowly pen might wait upon these four!
I bring my four times four, now meanly clad,
To do their homage unto yours, full glad:
Who for their age, their worth, and quality
Might seem of yours to claim precedency:
But by my humble hand thus rudely penned,
They are your bounden handmaids to attend.
These same are they from whom we being have;
These are of all the life, the nurse, the grave;
These are the hot, the cold, the moist, the dry,
That sink, that swim, that fill, that upwards fly;

Of these consist our bodies, clothes, and food,
The world, the useful, hurtful, and the good.
Sweet harmony they keep, yet jar ofttimes —
Their discord doth appear by these harsh rhymes.
Yours did contest for wealth, for arts, for age;
My first do show their good, and then their rage.
My other Fours do intermixéd tell
Each other's faults, and where themselves excel;
How hot and dry contend with moist and cold,
How air and earth no correspondence hold,
And yet, in equal tempers, how they agree,
How divers natures make one unity.
Something of all, though mean, I did intend,
But feared you 'd judge Du Bartas was my friend.
I honor him, but dare not wear his wealth.
My goods are true, though poor; I love no stealth;
But if I did I durst not send them you,
Who must reward a thief but with his due.
I shall not need mine innocence to clear:
These ragged lines will do it when they appear.
On what they are, your mild aspect I crave;
Accept my best, my worst vouchsafe a grave.

From her that to yourself more duty owes
Than water in the boundless ocean flows.

   March 20, 1642.
                   ANNE BRADSTREET.

THE PROLOGUE.

To sing of wars, of captains, and of kings,
Of cities founded, commonwealths begun,
For my mean pen are too superior things:
Or how they all, or each, their dates have run;
Let poets and historians set these forth,
My obscure lines shall not so dim their worth.

But when my wondering eyes and envious heart
Great Bartas' sugared lines do but read o'er,
Fool I do grudge the Muses did not part
'Twixt him and me that overfluent store;—
A Bartas can do what a Bartas will,
But simple I according to my skill.

From school-boys' tongues no rhetoric we expect,
Nor yet a sweet consort from broken strings,
Nor perfect beauty where 's a main defect:
My foolish, broken, blemished Muse so sings;
And this to mend, alas, no art is able,
'Cause nature made it so, irreparable.

Nor can I, like that fluent, sweet-tongued Greek
Who lisped at first, in future times speak plain;
By art he gladly found what he did seek —
A full requital of his striving pain.

Art can do much, but this maxim 's most sure:
A weak or wounded brain admits no cure.

I am obnoxious to each carping tongue
Who says my hand a needle better fits.
A poet's pen all scorn I should thus wrong;
For such despite they cast on female wits,
If what I do prove well, it won't advance —
They 'll say it 's stolen, or else it was by chance.

But sure the antique Greeks were far more mild,
Else of our sex why feignéd they those Nine,
And Poesy made Calliope's own child?
So 'mongst the rest they placed the Arts Divine.
But this weak knot they will full soon untie —
The Greeks did naught but play the fools and lie.

Let Greeks be Greeks, and women what they are.
Men have precedency, and still excel.
It is but vain unjustly to wage war:
Men can do best, and women know it well.
Preëminence in all and each is yours —
Yet grant some small acknowledgment of ours.

And oh, ye high flown quills that soar the skies,
And ever with your prey still catch your praise,
If e'er you deign these lowly lines your eyes,
Give thyme or parsley wreath; I ask no bays.
This mean and unrefinéd ore of mine
Will make your glistering gold but more to shine.

THE FOUR ELEMENTS.

The Fire, Air, Earth, and Water did contest
Which was the strongest, noblest, and the best;
Who was of greatest use and mightiest force.
In placid terms they thought now to discourse,
That in due order each her turn should speak.
But enmity this amity did break:
All would be chief, and all scorned to be under;
Whence issued winds and rains, lightning and thunder;
The quaking earth did groan, the sky looked black,
The fire the forcéd air in sunder crack;
The sea did threat the heavens, the heavens the earth;
All lookéd like a chaos, or new birth.
Firé broiléd earth, and scorchéd earth it choked;
Both, by their darings, water so provoked
That roaring in it came, and with its source
Soon made the combatants abate their force.
The rumbling, hissing, puffing, was so great
The world's confusion it did seem to threat;
Till gentle Air contention so abated
That betwixt hot and cold she arbitrated.

The others' difference, being less, did cease,
All storms now laid, and they in perfect peace.
That Fire should first begin the rest consent,
The noblest and most active element.

FIRE.

"What is my worth both ye and all men know.
In little time I can but little show.
But what I am, let learned Grecians say;
What I can do, well-skilled mechanics may;
The benefit all living by me find,
All sorts of artists here declare your mind.
What tool was ever framed but by my might?
Ye martialists, what weapons for your fight,
To try your valor by, but it must feel
My force?—your sword, and gun, your lance of steel.
Your cannon's bootless, and your powder, too,
Without mine aid. Alas, what can they do—
The adverse wall's not shaked, the mine's not blown,
And in despite the city keeps her own.
But I with one granado or petard
Set ope those gates that 'fore so strong were barred.
Ye husbandmen, your coulters 're made by me,
Your hoes, your mattocks, and whate'er you see
Subdue the earth, and fit it for your grain,
That so it might in time requite your pain;
Though strong-limbed Vulcan forged it by his skill,
I made it flexible unto his will.

The Four Elements

Ye cooks, your kitchen implements I frame,
Your spits, pots, jacks, what else I need not name;
Your daily food I wholesome make; I warm
Your shrinking limbs, which winter's cold doth harm.
Ye Paracelsians, too, in vain 's your skill
In chemistry unless I help you still.
And you, philosophers, if e'er you made
A transmutation it was through mine aid.
Ye silversmiths, your ore I do refine;
What mingled lay with earth I cause to shine.
But let me leave these things; my flame aspires
To match on high with the celestial fires.
The sun an orb of fire was held of old;
Our sages now another tale have told.
But be he what they will, yet his aspect
A burning fiery heat we find reflect;
And of the self-same nature is with mine,
Cold sister Earth, no witness needs but thine:
How doth his warmth refresh thy frozen back,
And trim thee brave in green after thy black!
Both man and beast rejoice at his approach,
And birds do sing to see his glittering coach.
And though naught but salamanders live in fire,
And fly *pyrausta* called,—all else expire,—
Yet men and beasts, astronomers will tell,
Fixed in heavenly constellations dwell —
My planets of both sexes, whose degree
Poor heathen judged worthy a deity.

There's Orion, armed, attended by his dog;
The Theban, stout Alcides, with his club;
The valiant Perseus, who Medusa slew;
The horse that killed Bellerophon, then flew.
My crab, my scorpion, fishes, you may see,
The maid with balance, wain with horses three,
The ram, the bull, the lion, and the beagle,
The bear, the goat, the raven, and the eagle,
The crown, the whale, the archer, Berenice's hair,
The hydra, dolphin, boys that water bear;
Nay, more than these, rivers 'mongst stars are found —
Eridanus, where Phaethon was drowned.
Their magnitude and height should I recount,
My story to a volume would amount.
Out of a multitude these few I touch;
Your wisdom out of little gather much.
I'll here let pass my choler, cause of wars;
And influence of divers of those stars,
When in conjunction with the sun, do more
Augment his heat which was too hot before.
The summer ripening season I do claim;
And man from thirty unto fifty frame.
Of old, when sacrifices were divine,
I of acceptance was the holy sign.
'Mong all my wonders which I might recount,
There's none more strange than Ætna's sulph'ry mount;
The choking flames that from Vesuvius flew
The over-curious Second Pliny slew,

The Four Elements

And with the ashes that it sometimes shed
Apulia's 'jacent parts were coveréd.
And though I be a servant to each man,
Yet, by my force, master my masters can.
What famous towns to cinders have I turned!
What lasting forts my kindled wrath hath burned!
The stately seats of mighty kings by me
In confused heaps of ashes may you see.
Where's Ninus' great walled town, and Troy of old,
Carthage, and hundred more in stories told?
Which when they could not be o'ercome by foes,
The army, through my help, victorious rose.
And stately London, our Great Britain's glory,
My raging flame did make a mournful story;—
But maugre all that I or foes could do,
That phenix from her bed is risen new.
Old sacred Zion, I demolished thee;
Low great Diana's temple was by me;
And more than bruitish Sodom for her lust,
With neighboring towns, I did consume to dust.
What shall I say of lightning and of thunder,
Which kings and mighty ones amaze with wonder,—
Which made a Cæsar (Rome's), the world's proud head,
Foolish Caligula, creep under his bed,—
Of meteors, ignes fatui, and the rest?
But to leave those to the wise I judge it best.
The rich I oft make poor, the strong I maim,
Not sparing life when I can take the same.

And, in a word, the world I shall consume,
And all therein, at that great day of doom;
Not before then shall cease my raging ire,
And then because no matter more for fire.
Now, sisters, pray proceed; each in your course,
As I, impart your usefulness and force."

EARTH.

The next in place Earth judged to be her due.
"Sister," quoth she, "I come not short of you;
In wealth and use I do surpass you all,
And Mother Earth of old men did me call,
Such is my fruitfulness — an epithet
Which none e'er gave, or you could claim, of right.
Among my praises this I count not least,
I am the original of man and beast.
To tell what sundry fruits my fat soil yields
In vineyards, gardens, orchards, and corn-fields,
Their kinds, their tastes, their colors, and their smells,
Would so pass time I could say nothing else;
The rich, the poor, wise, fool, and every sort,
Of these so common things can make report.
To tell you of my countries and my regions,
Soon would they pass not hundreds but legions;
My cities famous, rich, and populous,
Whose numbers now are grown innumerous.
I have not time to think of every part,
Yet let me name my Grecia, 't is my heart;

The Four Elements

For learning, arms, and arts I love it well,
But chiefly 'cause the Muses there did dwell.
I'll here skip o'er my mountains reaching sky,
Whether Pyrenean or the Alps, which lie
On either side the country of the Gauls,
Strong forts from Spanish and Italian brawls;
And huge great Taurus, longer than the rest,
Dividing great Armenia from the least;
And Hemus, whose steep sides none foot upon.
But farewell all for dear Mount Helicon;
And wondrous high Olympus, of such fame
That heaven itself was oft called by that name;
Parnassus sweet, I dote too much on thee,
Unless thou prove a better friend to me.
But I'll leap o'er these hills, not touch a dale,
Nor will I stay, no, not in Tempe vale.
I'll here let go my lions of Numidia,
My panthers and my leopards of Libya,
The behemoth, and rare found unicorn
(Poison's sure antidote lies in his horn),
And my hyena (imitates man's voice);
Out of great numbers I might pick my choice,
Thousands in woods and plains, both wild and tame.
But here or there, I list now none to name —
No, though the fawning dog did urge me sore
In his behalf to speak a word the more,
Whose trust and valor I might here commend,
But time's too short and precious so to spend.

But hark you, wealthy merchants, who for prize
Send forth your well-manned ships where sun doth rise:
After three years, when men and meat are spent,
My rich commodities pay double rent.
Ye Galenists, my drugs that come from thence
Do cure your patients, fill your purse with pence;
Besides the use of roots, of herbs, and plants
That with less cost near home supply your wants.
But, mariners, where got you ship and sail,
And oars to row, when both my sisters fail?
Your tackling, anchor, compass, too, is mine,
Which guides when sun nor moon nor stars do shine.
Ye mighty kings, who for your lasting fames
Built cities, monuments, called by your names,
Were those compiléd heaps of massy stones
That your ambition laid aught but my bones?
Ye greedy misers, who do dig for gold,
For gems, for silver, treasures which I hold,
Will not my goodly face your rage suffice
But you will see what in my bowels lies?
And ye artificers, all trades and sorts,
My bounty calls you forth to make reports
If aught you have to use, to wear, to eat,
But what I freely yield upon your sweat?
And choleric sister, thou, for all thine ire,
Well knowest my fuel must maintain thy fire;
As I ingenuously with thanks confess,
My cold thy fruitful heat doth crave no less

The Four Elements

But how my cold, dry temper works upon
The melancholy constitutiön,
How the autumnal season I do sway,
And how I force the grayhead to obey,
I should here make a short yet true narration,
But that thy method is mine imitation.
Now must I show mine adverse quality,
And how I oft work man's mortality.
He sometimes finds, maugre his toiling pain,
Thistles and thorns where he expected grain;
My sap to plants and trees I must not grant;
The vine, the olive, and the fig-tree want;
The corn and hay do fall before they 're mown;
And buds from fruitful trees as soon as blown.
Then dearth prevails; that nature to suffice,
The mother on her tender infant flies;
The husband knows no wife, nor father sons,
But to all outrages their hunger runs.
Dreadful examples soon I might produce,
But to such auditors 't were of no use.
Again, when delvers dare, in hope of gold,
To ope those veins of mine, audacious, bold,
While they thus in mine entrails love to dive,
Before they know they are interred alive.
Ye affrighted wights appalled, how do ye shake
When once you feel me, your foundation, quake?—
Because in the abyss of my dark womb
Your cities and your selves I oft entomb.

O dreadful sepulcher! that this is true,
Dathan and all his company well knew;
So did that Roman, far more stout than wise,
Burying himself alive for honor's prize;
And since fair Italy full sadly knows
What she hath lost by these remediless woes.
Again, what veins of poison in me lie;
Some kill outright, and some do stupefy—
Nay, into herbs and plants it sometimes creeps,
In heats, and colds, and gripes, and drowsy sleeps.
Thus I occasion death to man and beast
When food they seek and harm mistrust the least.
Much might I say of the hot Libyan sand
Which rise like tumbling billows on the land
Wherein Cambyses' army was o'erthrown
(But, windy sister, 't was when you have blown).
I 'll say no more; but this thing add I must:
Remember, sons, your mold is of my dust;
And after death, whether interred or burned,
As earth at first, so into earth returned."

WATER.

Scarce Earth had done, but the angry Water moved.
"Sister," quoth she, "it had full well behooved,
Among your boastings, to have praiséd me,
Cause of your fruitfulness, as you shall see.
This, your neglect, shows your ingratitude,
And how your subtilty would men delude.

The Four Elements

Not one of us, all knows, that's like to thee,
Ever, in craving from the other three.
But thou art bound to me above the rest,
Who am thy drink, thy blood, thy sap, and best.
If I withhold, what art thou? Dead, dry lump,
Thou bearest nor grass, nor plant, nor tree, nor stump.
Thy extreme thirst is moistened by my love
With springs below and showers from above,
Or else thy sun-burned face and gaping chops
Complain to the heavens, if I withhold my drops.
Thy bear, thy tiger, and thy lion stout,
When I am gone their fierceness none needs doubt;
Thy camel hath no strength, thy bull no force,
Nor mettle's found in the courageous horse;
Hinds leave their calves, the elephant the fens,
The wolves and savage beasts forsake their dens;
The lofty eagle and the stork fly low;
The peacock and the ostrich share in woe;
The pine, the cedar, yea, and Daphne's tree
Do cease to flourish in this misery.
Man wants his bread and wine, and pleasant fruits;
He knows such sweets lie not in Earth's dry roots,
Then seeks me out, in river and in well,
His deadly malady I might expel.
If I supply, his heart and veins rejoice;
If not, soon ends his life, as did his voice.
That this is true, Earth, thou canst not deny.
I call thine Egypt this to verify,

Which, by my fatting Nile, doth yield such store
That she can spare when nations round are poor;
When I run low, and not o'erflow her brinks,
To meet with want each woeful man bethinks.
And such I am, in rivers, showers, and springs.
But what's the wealth that my rich ocean brings?—
Fishes so numberless I there do hold,
If thou shouldst buy it would exhaust thy gold.
There lives the oily whale, whom all men know,—
Such wealth, but not such like, Earth, thou mayst show,—
The dolphin, loving music, Arion's friend,
The witty barbel, whose craft doth her commend,
With thousands more which now I list not name,
Thy silence of thy beasts doth cause the same.
My pearls that dangle at thy darlings' ears
Not thou, but shell-fish, yield, as Pliny clears.
Was ever gem so rich found in thy trunk
As Egypt's wanton Cleopatra drunk?
Or hast thou any color can come nigh
The Roman purple, double Tyrian dye?—
Which Cæsar's consuls, tribunes, all adorn,
For it to search my waves they thought no scorn.
Thy gallant, rich, perfuming ambergris
I lightly cast ashore as frothy fleece;
With rolling grains of purest massy gold,
Which Spain's Americas do gladly hold.
Earth, thou hast not more countries, vales, and mounds
Than I have fountains, rivers, lakes, and ponds:

The Four Elements

My sundry seas, Black, White, and Adriatic,
Ionian, Baltic, and the vast Atlantic,
Ægean, Caspian, golden rivers five,
Asphaltites Lake, where naught remains alive;—
But I should go beyond thee in my boasts
If I should name more seas than thou hast coasts.
And be thy mountains e'er so high and steep,
I soon can match them with my seas as deep.
To speak of kinds of waters I neglect—
My divers fountains, and their strange effect;
My wholesome baths, together with their cures;
My water sirens, with their guileful lures;
The uncertain cause of certain ebbs and flows,
Which wondering Aristotle's wit ne'er knows.
Nor will I speak of waters made by art,
Which can to life restore a fainting heart;
Nor fruitful dews; nor drops distilled from eyes,
Which pity move, and oft deceive the wise;
Nor yet of salt and sugar, sweet and smart—
Both, when we list, to water we convert.
Alas, thy ships and oars could do no good
Did they but want my ocean and my flood.
The wary merchant on his weary beast
Transfers his goods from south to north and east,
Unless I ease his toil and do transport
The wealthy freight unto his wishéd port.
These be my benefits, which may suffice.
I now must show what ill there in me lies.

The phlegmy constitution I uphold;
All humors, tumors, which are bred of cold.
O'er childhood and o'er winter I bear sway,
And Luna for my regent I obey.
As I with showers ofttimes refresh the earth,
So oft in my excess I cause a dearth,
And with abundant wet so cool the ground,
By adding cold to cold, no fruit proves sound.
The farmer and the grasier do complain
Of rotten sheep, lean kine, and mildewed grain.
And with my wasting floods and roaring torrent
Their cattle, hay, and corn I sweep down current.
Nay, many times my ocean breaks his bounds,
And with astonishment the world confounds,
And swallows countries up, ne'er seen again,
And that an island makes which once was main.
Thus Britain fair, 't is thought, was cut from France;
Sicily from Italy by the like chance;
And but one land was Africa and Spain
Until proud Gibraltar did make them twain.
Some say I swallowed up (sure 't is a notion)
A mighty country in the Atlantic Ocean.
I need not say much of my hail and snow,
My ice and extreme cold, which all men know;
Whereof the first so ominous I rained
That Israel's enemies therewith were brained;
And of my chilling snows such plenty be
That Caucasus' high mounts are seldom free.

Mine ice doth glaze Europe's great rivers o'er;
Till sun release, their ships can sail no more.
All know that inundations I have made,
Wherein not men, but mountains, seemed to wade:
As when Achaia all under water stood,
That for two hundred years it ne'er proved good;
Deucalion's great deluge, with many more.
But these are trifles to the flood of Noah;
Then wholly perished earth's ignoble race,
And to this day impairs her beauteous face.
That after times shall never feel like woe,
Her confirmed sons behold my colored bow.
Much might I say of wrecks; but that I'll spare,
And now give place unto our sister Air."

AIR.

"Content," quoth Air, "to speak the last of you,
Yet am not ignorant first was my due.
I do suppose you'll yield, without control,
I am the breath of every living soul.
Mortals, what one of you that loves not me
Abundantly more than my sisters three?
And though you love Fire, Earth, and Water well,
Yet Air beyond all these you know to excel.
I ask the man condemned, that's near his death,
How gladly should his gold purchase his breath;
And all the wealth that ever earth did give,
How freely should it go, so he might live.

No, Earth, thy witching trash were all but vain
If my pure air thy sons did not sustain.
The famished, thirsty man that craves supply,
His moving reason is, Give, lest I die,
So loth he is to go, though nature's spent,
To bid adieu to his dear element.
Nay, what are words, which do reveal the mind?—
Speak who or what they will, they are but wind.
Your drums', your trumpets', and your organs' sound,
What is it but forcéd air which doth rebound?
And such are echoes, and report of the gun
That tells afar the exploit which it hath done.
Your songs and pleasant tunes, they are the same;
And so the notes which nightingales do frame.
Ye forging smiths, if bellows once were gone
Your red-hot work more coldly would go on.
Ye mariners, 't is I that fills your sails,
And speeds you to your port with wishéd gales.
When burning heat doth cause you faint, I cool;
And when I smile, your ocean's like a pool.
I help to ripe the corn, I turn the mill,
And with myself I every vacuum fill.
The ruddy, sweet sanguine is like to air,
And youth and spring sages to me compare.
My moist, hot nature is so purely thin,
No place so subtilely made but I get in.
I grow more pure and pure as I mount higher,
And when I'm throughly rarefied, turn fire.

The Four Elements

So, when I am condensed, I turn to water,
Which may be done by holding down my vapor;
Thus I another body can assume,
And in a trice my own nature resume.
Some for this cause of late have been so bold
Me for no element longer to hold.
Let such suspend their thoughts, and silent be,
For all philosophers make one of me;
And what those sages either spake or writ
Is more authentic than our modern wit.
Next, of my fowls such multitudes there are,
Earth's beasts and Water's fish scarce can compare —
The ostrich with her plumes, the eagle with her eyen,
The phenix, too, if any be, are mine;
The stork, the crane, the partridge, and the pheasant,
The thrush, the wren, the lark,—a prey to the peasant,—
With thousands more which now I may omit
Without impeachment to my tale or wit.
As my fresh air preserves all things in life,
So, when corrupt, mortality is rife:
Then fevers, purples, pox, and pestilence,
With divers more, work deadly consequence;
Whereof such multitudes have died and fled,
The living scarce had power to bury the dead.
Yea, so contagious countries have we known,
That birds have not 'scaped death as they have flown;
Of murrain, cattle numberless did fall;
Men feared destruction epidemical.

Then of my tempests felt at sea and land,
Which neither ships nor houses could withstand,
What woeful wrecks I 've made may well appear,
If naught were known but that before Algier,
Where famous Charles the Fifth more loss sustained
Than in the long hot war which Milan gained.
Again, what furious storms and hurricanoes
Know western isles, as Christopher's, Barbadoes,
Where neither houses, trees, nor plants I spare,
But some fall down, and some fly up with air.
Earthquakes so hurtful, and so feared of all,
Imprisoned I am the original.
Then what prodigious sights I sometimes show,
As battles pitched in the air, as countries know;
Their joining, fighting, forcing, and retreat,
That earth appears in heaven, oh, wonder great!
Sometimes red flaming swords and blazing stars,
Portentous signs of famines, plagues, and wars,
Which make the mighty monarchs fear their fates
By death, or great mutation of their states.
I have said less than did my sisters three;
But what 's their wrath or force, the same 's in me.
To add to all I 've said was my intent,
But dare not go beyond my element."

OF THE FOUR HUMORS IN MAN'S CONSTITUTION.

The former four now ending their discourse,
Ceasing to vaunt their good, or threat their force,
Lo, other four step up, crave leave to show
The native qualities that from them flow.
But first they wisely showed their high descent,
Each eldest daughter to each element:
Choler was owned by Fire, and Blood by Air;
Earth knew her black swarth child, Water her fair.
All having made obeisance to each mother,
Had leave to speak, succeeding one the other.
But 'mongst themselves they were at variance
Which of the four should have predominance.
Choler first hotly claimed right by her mother,
Who had precedency of all the other;
But Sanguine did disdain what she required,
Pleading herself was most of all desired.
Proud Melancholy, more envious than the rest,
The second, third, or last could not digest;

She was the silentest of all the four;
Her wisdom spake not much, but thought the more.
Mild Phlegm did not contest for chiefest place,
Only she craved to have a vacant space.
Well, thus they parle and chide; but, to be brief,
Or will they nill they Choler will be chief.
They, seeing her impetuosity,
At present yielded to necessity.

CHOLER.

"To show my high descent and pedigree
Yourselves would judge but vain prolixity.
It is acknowledgéd from whence I came;
It shall suffice to show you what I am —
Myself and mother one, as you shall see,
But she in greater, I in less, degree.
We both once masculines, the world doth know,
Now feminines awhile, for love we owe
Unto your sisterhood, which makes us render
Our noble selves in a less noble gender.
Though under fire we comprehend all heat,
Yet man for choler is the proper seat;
I in his heart erect my regal throne,
Where monarch-like I play and sway alone.
Yet many times, unto my great disgrace,
One of yourselves are my compeers in place,
Where if your rule prove once predominant,
The man proves boyish, sottish, ignorant;

The Four Humors

But if you yield subservience unto me,
I make a man a man in the highest degree.
Be he a soldier, I more fence his heart
Than iron corslet 'gainst a sword or dart.
What makes him face his foe without appal,
To storm a breach, or scale a city wall;
In dangers to account himself more sure
Than timorous hares whom castles do immure?
Have you not heard of worthies, demi-gods?
'Twixt them and others what is it makes the odds
But valor? Whence comes that? From none of you.
Nay, milksops, at such brunts you look but blue.
Here's sister Ruddy, worth the other two,
Who much will talk, but little dares she do,
Unless to court and claw, to dice and drink;
And there she will outbid us all, I think.
She loves a fiddle better than a drum;
A chamber well; in field she dares not come.
She'll ride a horse as bravely as the best,
And break a staff, provided be in jest;
But shuns to look on wounds, and blood that's spilt.
She loves her sword only because it's gilt.
Then here's our sad black sister, worse than you;
She'll neither say she will, nor will she do,
But, peevish malcontent, she musing sits,
And by misprision's like to lose her wits.
If great persuasions cause her meet her foe,
In her dull resolution she's so slow

To march her pace to some is greater pain
Than by a quick encounter to be slain.
But be she beaten, she 'll not run away;
She 'll first advise if it be not best to stay.
Now let 's give cold white sister Phlegm her right —
So loving unto all, she scorns to fight;
If any threaten her, she 'll in a trice
Convert from water to congealéd ice;
Her teeth will chatter, dead and wan 's her face,
And 'fore she be assaulted quits the place.
She dares not challenge if I speak amiss,
Nor hath she wit or heat to blush at this.
Here 's three of you all see now what you are;
Then yield to me preëminence in war.
Again, who fits for learning, science, arts?
Who rarefies the intellectual parts,
From whence fine spirits flow, and witty notions?
But 't is not from our dull slow sister's motions,
Nor, sister Sanguine, from thy moderate heat.
Poor spirits the liver breeds, which is thy seat.
What comes from thence my heat refines the same,
And through the arteries sends it o'er the frame;
The vital spirits they 're called, and well they may,
For when they fail man turns unto his clay.
The animal I claim as well as these —
The nerves should I not warm, soon would they freeze.
But Phlegm herself is now provoked at this.
She thinks I never shot so far amiss;

The Four Humors

The brain she challengeth, the head's her seat.
But know it's a foolish brain that wanteth heat;
My absence proves it plain — her wit then flies
Out at her nose, or melteth at her eyes.
Oh, who would miss this influence of thine
To be distilled, a drop on every line!
Alas, thou hast no spirits; thy company
Will feed a dropsy or a tympany,
The palsy, gout, or cramp, or some such dolor.
Thou wast not made for soldier or for scholar.
Of greasy paunch and bloated cheeks go vaunt;
But a good head from these are dissonant.
But, Melancholy, wouldst have this glory thine?
Thou sayest thy wits are staid, subtile, and fine;
'T is true, when I am midwife to thy birth
Thyself's as dull as is thy mother Earth.
Thou canst not claim the liver, head, nor heart,
Yet hast the seat assigned, a goodly part —
The sink of all us three, the hateful spleen;
Of that black region nature made thee queen,
Where pain and sore obstruction thou dost work,
Where envy, malice, thy companions, lurk.
If once thou 'rt great, what follows thereupon
But bodies wasting and destruction?
So base thou art that baser cannot be,
The excrement adustion of me.
But I am weary to dilate your shame;
Nor is it my pleasure thus to blur your name,

Only to raise my honor to the skies,
As objects best appear by contraries.
But arms and arts I claim, and higher things,
The princely qualities befitting kings,
Whose profound heads I line with policies:
They're held for oracles, they are so wise;
Their wrathful looks are death, their words are laws;
Their courage it foe, friend, and subject awes.
But one of you would make a worthy king
Like our sixth Henry, that same virtuous thing
That, when a varlet struck him o'er the side,
'Forsooth, you are to blame,' he grave replied.
Take choler from a prince, what is he more
Than a dead lion, by beasts triumphéd o'er?
Again, you know how I act every part
By the influence I still send from the heart;
It's nor your muscles, nerves, nor this, nor that
Does aught without my lively heat, that's flat.
Nay, the stomach, magazine to all the rest,
Without my boiling heat cannot digest.
And yet, to make my greatness still more great,
What differences the sex but only heat?
And one thing more, to close up my narration,
Of all that lives I cause the propagation.
I have been sparing what I might have said.
I love no boasting; that's but children's trade.
To what you now shall say I will attend,
And to your weakness gently condescend."

BLOOD.

"Good sisters, give me leave, as is my place,
To vent my grief and wipe off my disgrace.
Yourselves may plead your wrongs are no whit less—
Your patience more than mine I must confess.
Did ever sober tongue such language speak,
Or honesty such ties unfriendly break?
Dost know thyself so well, us so amiss?
Is it arrogance or folly causeth this?
I'll only show the wrong thou'st done to me,
Then let my sisters right their injury.
To pay with railings is not mine intent,
But to evince the truth by argument.
I will analyze this thy proud relation,
So full of boasting and prevarication;
Thy foolish incongruities I'll show,
So walk thee till thou'rt cold, then let thee go.
There is no soldier but thyself, thou sayest;
No valor upon earth, but what thou hast.
Thy silly provocations I despise,
And leave it to all to judge where valor lies.
No pattern, nor no patron, will I bring
But David, Judah's most heroic king,
Whose glorious deeds in arms the world can tell,
A rosy-cheeked musician thou knowest well;
He knew well how to handle sword and harp,
And how to strike full sweet, as well as sharp.

Thou laughest at me for loving merriment,
And scornest all knightly sports at tournament.
Thou sayest I love my sword because it's gilt;
But know I love the blade more than the hilt,
Yet do abhor such temerarious deeds
As thy unbridled barbarous choler breeds.
Thy rudeness counts good manners vanity,
And real compliments base flattery.
For drink, which of us twain likes it the best
I'll go no farther than thy nose for test.
Thy other scoffs, not worthy of reply,
Shall vanish as of no validity.
Of thy black calumnies this is but part,
But now I'll show what soldiër thou art.
And though thou'st used me with opprobrious spite,
My ingenuity must give thee right.
Thy choler is but rage when 't is most pure,
But useful when a mixture can endure.
As with thy mother Fire, so 't is with thee —
The best of all the four when they agree;
But let her leave the rest, then I presume
Both them and all things else she would consume.
Whilst us for thine associates thou takest,
A soldier most complete in all points makest;
But when thou scornest to take the help we lend,
Thou art a fury or infernal fiend.
Witness the execrable deeds thou 'st done,
Nor sparing sex nor age, nor sire nor son.

The Four Humors

To satisfy thy pride and cruelty
Thou oft hast broke bounds of humanity.
Nay, should I tell, thou wouldst count me no blab,
How often for the lie thou 'st given the stab.
To take the wall 's a sin of so high rate
That naught but death the same may expiate.
To cross thy will, a challenge doth deserve;
So sheddest that blood thou 'rt bounden to preserve.
Wilt thou this valor, courage, manhood, call?
No; know 't is pride most diabolical.
If murders be thy glory, 't is no less.
I 'll not envy thy feats nor happiness.
But if in fitting time and place 'gainst foes
For country's good thy life thou darest expose,
Be dangers ne'er so high, and courage great,
I 'll praise that prowess, fury, choler, heat.
But such thou never art when all alone,
Yet such when we all four are joined in one.
And when such thou art, even such are we,
The friendly coadjutors still of thee.
Nextly, the spirits thou dost wholly claim,
Which natural, vital, animal, we name.
To play philosopher I have no list,
Nor yet physician, nor anatomist;
For acting these I have no will nor art,
Yet shall with equity give thee thy part.
For natural, thou dost not much contest;
For there is none, thou sayest, if some, not best.

That there are some, and best, I dare aver,
Of greatest use, if reason do not err.
What is there living which does not first derive
His life, now animal, from vegetive?
If thou givest life, I give the nourishment;
Thine without mine is not, 't is evident.
But I, without thy help, can give a growth,
As plants, trees, and small embryon knoweth.
And if vital spirits do flow from thee,
I am as sure the natural from me.
Be thine the nobler, which I grant, yet mine
Shall justly claim priority of thine.
I am the fountain which thy cistern fills
Through warm blue conduits of my venial rills.
What hath the heart but what 's sent from the liver?
If thou 'rt the taker, I must be the giver.
Then never boast of what thou dost receive,
For of such glory I shall thee bereave.
But why the heart should be usurped by thee
I must confess seems something strange to me.
The spirits through thy heat made perfect are,
But the material 's none of thine, that 's clear;
Their wondrous mixture is of blood and air—
The first, myself; second, my mother fair.
But I 'll not force retorts, nor do thee wrong;
Thy fiery yellow froth is mixed among.
Challenge not all 'cause part we do allow;
Thou knowest I 've there to do as well as thou.

The Four Humors

But thou wilt say I deal unequally.
There lives the irascible faculty
Which, without all dispute, is Choler's own;
Besides, the vehement heat, only there known,
Can be imputed unto none but Fire,
Which is thyself, thy mother, and thy sire.
That this is true I easily can assent,
If still you take along my aliment,
And let me be your partner, which is due;
So shall I give the dignity to you.
Again, stomach's concoction thou dost claim,
But by what right nor dost nor canst thou name,
Unless, as heat, it be thy faculty,
And so thou challengest her property.
The help she needs the loving liver lends,
Who the benefit of the whole ever intends.
To meddle further I shall be but shent,
The rest to our sisters is more pertinent;
Your slanders, thus refuted, take no place,
Nor what you've said doth argue my disgrace.
Now through your leaves some little time I'll spend
My worth in humble manner to commend.
This hot, moist, nutritive humor of mine,
When 't is untaint, pure, and most genuine,
Shall chiefly take the place, as is my due,
Without the least indignity to you.
Of all your qualities I do partake,
And what you single are, the whole I make.

Your hot, moist, cold, dry natures are but four.
I moderately am all; what need I more?
As thus, if hot, then dry; if moist, then cold.
If this you can't disprove, then all I hold.
My virtues hid, I 've let you dimly see
My sweet complexion proves the verity.
This scarlet dye 's a badge of what 's within,
One touch thereof so beautifies the skin.
Nay, could I be from all your tangs but pure,
Man's life to boundless time might still endure.
But here one thrusts her heat, where it 's not required;
So, suddenly, the body all is fired,
And of the calm, sweet temper quite bereft,
Which makes the mansion by the soul soon left.
So Melancholy seizes on a man,
With her uncheerful visage, swarth and wan;
The body dries, the mind sublime doth smother,
And turns him to the womb of his earthy mother.
And Phlegm likewise can show her cruel art,
With cold distempers to pain every part;
The lungs she rots, the body wears away,
As if she 'd leave no flesh to turn to clay.
Her languishing diseases, though not quick,
At length, alas, demolish the fabric.
All to prevent, this curious care I take:
In the last concoction segregation make
Of all the perverse humors from mine own.
The bitter Choler, most malignant known,

The Four Humors

I turn into her cell close by my side;
The Melancholy to the spleen to abide;
Likewise the whey, some use I in the veins,
The overplus I send unto the reins.
But yet, for all my toil, my care, and skill,
It 's doomed, by an irrevocable will,
That my intents should meet with interruption,
That mortal man might turn to his corruption.
I might here show the nobleness of mind
Of such as to the sanguine are inclined;
They 're liberal, pleasant, kind, and courteous,
And, like the liver, all benignious.
For arts and sciences they are the fittest,
And, maugre Choler, still they are the wittiest,
With an ingenious working fantasy,
A most voluminous large memory,
And nothing wanting but solidity.
But why, alas, thus tedious should I be?
Thousand examples you may daily see.
If time I have transgressed, and been too long,
Yet could not be more brief without much wrong.
I 've scarce wiped off the spots proud Choler cast,
Such venom lies in words, though but a blast.
No brags I 've used; to you I dare appeal,
If modesty my worth do not conceal.
I 've used no bitterness, nor taxed your name.
As I to you, to me do ye the same."

MELANCHOLY.

"He that with two assailants hath to do
Had need be arméd well, and active too —
Especially when friendship is pretended;
That blow's most deadly where it is intended.
Though Choler rage and rail, I'll not do so;
The tongue's no weapon to assault a foe.
But sith we fight with words, we might be kind
To spare ourselves and beat the whistling wind.
Fair rosy sister, so mightest thou 'scape free.
(I'll flatter for a time as thou didst me;
But when the first offender I have laid,
Thy soothing girds shall fully be repaid.)
But, Choler, be thou cooled or chafed, I'll venture,
And in contention's lists now justly enter.
What moved thee thus to vilify my name,
Not past all reason, but, in truth, all shame?
Thy fiery spirit shall bear away this prize;
To play such furious pranks I am too wise.
If in a soldier rashness be so precious,
Know in a general it is most pernicious.
Nature doth teach to shield the head from harm;
The blow that's aimed thereat is latched by the arm.
When in battalia my foes I face,
I then command proud Choler stand thy place,
To use thy sword, thy courage, and thy art
There to defend myself, thy better part.

The Four Humors 51

This wariness count not for cowardice;
He is not truly valiant that 's not wise.
It 's no less glory to defend a town
Than by assault to gain one not our own.
And if Marcellus bold be called Rome's sword,
Wise Fabius is her buckler, all accord.
And if thy haste my slowness should not temper,
'T were but a mad, irregular distemper.
Enough of that by our sisters heretofore.
I 'll come to that which wounds me somewhat more.
Of learning, policy, thou wouldst bereave me,
But not thine ignorance shall thus deceive me.
What greater clerk or politician lives
Than he whose brain a touch my humor gives?
What is too hot my coldness doth abate,
What 's diffluent I do consolidate.
If I be partial judged, or thought to err,
The melancholy snake shall it aver,
Whose cold dry head more subtility doth yield
Than all the huge beasts of the fertile field.
Again, thou dost confine me to the spleen,
As of that only part I were the queen.
Let me as well make thy precincts the gall,
So prison thee within that bladder small.
Reduce the man to his principles, then see
If I have not more part than all you three.
What is within, without, of theirs or thine,
Yet time and age shall soon declare it mine.

When death doth seize the man, your stock is lost;
When you poor bankrupts prove, then have I most.
You 'll say, here none shall e'er disturb my right;
You, high born, from that lump then take your flight.
Then who's man's friend, when life and all forsakes?
His mother, mine, him to her womb retakes:
Thus he is ours; his portion is the grave.
But while he lives, I 'll show what part I have.
And first, the firm dry bones I justly claim,
The strong foundation of the stately frame.
Likewise the useful spleen, though not the best,
Yet is a bowel called well as the rest;
The liver, stomach, owe their thanks of right:
The first it drains, of the last quicks appetite.
Laughter (though thou say malice) flows from hence —
These two in one cannot have residence.
But thou most grossly dost mistake to think
The spleen for all you three was made a sink.
Of all the rest thou 'st nothing there to do;
But if thou hast, that malice is from you.
Again, you often touch my swarthy hue.
That black is black, and I am black, is true,
But yet more comely far, I dare avow,
Than is thy torrid nose or brazen brow.
But that which shows how high your spite is bent
Is charging me to be thy excrement.
Thy loathsome imputation I defy.
So plain a slander needeth no reply.

The Four Humors

When by thy heat thou 'st baked thyself to crust,
And so art called 'black' Choler, or adust,
Thou, witless, thinkest that I am thy excretion,
So mean thou art in art as in discretion.
But by your leave I 'll let your greatness see
What officer thou art to us all three —
The kitchen drudge, the cleanser of the sinks,
That casts out all that man e'er eats or drinks.
If any doubt the truth whence this should come,
Show them thy passage to the duodenum;
Thy biting quality still irritates,
Till filth and thee nature exonerates.
If there thou 'rt stopped, to the liver thou dost turn in,
And thence with jaundice saffron all the skin.
No further time I 'll spend in confutation;
I trust I 've cleared your slanderous imputation.
I now speak unto all, no more to one;
Pray hear, admire, and learn instructiōn.
My virtues yours surpass without compare:
The first my constancy, that jewel rare.
Choler 's too rash this golden gift to hold,
And Sanguine is more fickle manifold; —
Here, there, her restless thoughts do ever fly,
Constant in nothing but unconstancy.
And what Phlegm is we know, like to her mother;
Unstable is the one, and so the other.
With me is noble patience also found.
Impatient Choler loveth not the sound.

What Sanguine is, she doth not heed nor care,
Now up, now down, transported like the air.
Phlegm's patient because her nature's tame.
But I by virtue do acquire the same.
My temperance, chastity, are eminent;
But these with you are seldom resident.
Now could I stain my ruddy sister's face
With deeper red, to show you her disgrace.
But rather I with silence veil her shame
Than cause her blush while I relate the same.
Nor are ye free from this enormity,
Although she bear the greatest obloquy.
My prudence, judgment, I might now reveal;
But wisdom 't is my wisdom to conceal.
Unto diseases not inclined as you,
Nor cold nor hot, ague nor pleurisy,
Nor cough nor quinsy, nor the burning fever,
I rarely feel to act his fierce endeavor.
My sickness in conceit chiefly doth lie;
What I imagine, that's my malady.
Chimeras strange are in my fantasy,
And things that never were, nor shall I see.
I love not talk; reason lies not in length,
Nor multitude of words argues our strength.
I've done. Pray, sister Phlegm, proceed in course.
We shall expect much sound, but little force."

PHLEGM.

"Patient I am, patient I'd need to be,
To bear with the injurious taunts of three.
Though wit I want, and anger I have less,
Enough of both my wrongs now to express.
I've not forgot how bitter Choler spake,
Nor how her gall on me she causeless brake;
Nor wonder 't was, for hatred there's not small
Where opposition is diametrical.
To what is truth I freely will assent,
Although my name do suffer detriment;
What's slanderous, repel; doubtful, dispute;
And when I've nothing left to say, be mute.
Valor I want; no soldier am, 't is true —
I'll leave that manly property to you;
I love no thundering guns, nor bloody wars;
My polished skin was not ordained for scars.
But though the pitchéd field I've ever fled,
At home the conquerors have conqueréd.
Nay, I could tell you what's more true than meet,
That kings have laid their scepters at my feet:
When sister Sanguine paints my ivory face,
The monarchs bend and sue but for my grace;
My lily-white, when joinéd with her red,
Princes hath slaved, and captains captivéd.
Country with country, Greece with Asia, fights,
Sixty-nine princes, all stout hero knights,

Under Troy's walls ten years will wear away
Rather than lose one beauteous Helena.
But 't were as vain to prove this truth of mine
As at noon-day to tell the sun doth shine.
Next difference that 'twixt us twain doth lie,
Who doth possess the brain, or thou or I?
Shame forced thee say the matter that was mine,
But the spirits by which it acts are thine.
Thou speakest truth, and I can say no less;
Thy heat doth much, I candidly confess.
Yet without ostentation I may say
I do as much for thee another way;
And though I grant thou art my helper here,
No debtor I because it 's paid elsewhere.
With all your flourishes, now, sisters three,
Who is it that dare, or can, compare with me?
My excellences are so great, so many,
I am confounded 'fore I speak of any.
The brain 's the noblest member, all allow;
Its form and situation will avow;
Its ventricles, membranes, and wondrous net
Galen, Hippocrates, drive to a set.
That divine offspring, the immortal soul,
Though it in all and every part be whole,
Within this stately place of eminence
Doth doubtless keep its mighty residence.
And surely the soul sensitive here lives,
Which life and motion to each creature gives.

The Four Humors

The conjugation of the parts to the brain
Doth show hence flow the powers which they retain:
Within this high-built citadel doth lie
The reason, fancy, and the memory;
The faculty of speech doth here abide;
The spirits animal from hence do slide;
The five most noble senses here do dwell —
Of three it's hard to say which doth excel.
This point now to discuss 'longs not to me;
I'll touch the sight, greatest wonder of the three.
The optic nerve, coats, humors, all are mine,
The watery, glassy, and the crystalline.
O mixture strange! O color colorless!
Thy perfect temperament who can express?
He was no fool who thought the soul lay there
Whence her affections, passions, speak so clear.
O good, O bad, O true, O traitorous eyes,
What wonderment within your balls there lies!
Of all the senses sight shall be the queen.
Yet some may wish, oh, had mine eyes ne'er seen!
Mine likewise is the marrow of the back,
Which runs through all the spondles of the rack;
It is the substitute of the royal brain;
All nerves, except seven pair, to it retain,
And the strong ligaments from hence arise
Which, joint to joint, the entire body ties.
Some other parts there issue from the brain,
Whose worth and use to tell I must refrain;

Some curious learned Crooke may these reveal,
But modesty hath charged me to conceal.
Here's my epitome of excellence,
For what's the brain's is mine by consequence.
A foolish brain, quoth Choler, wanting heat!
But a mad one, say I, where 't is too great.
Phrensy is worse than folly; one would more glad
With a tame fool converse than with a mad.
For learning, then, my brain is not the fittest,
Nor will I yield that Choler is the wittiest;
Thy judgment is unsafe, thy fancy little,
For memory the sand is not more brittle.
Again, none's fit for kingly state but thou!
If tyrants be the best, I'll it allow;
But if love be as requisite as fear,
Then thou and I must make a mixture here.
Well, to be brief, I hope now Choler's laid,
And I'll pass by what sister Sanguine said.
To Melancholy I'll make no reply;
The worst she said was instability
And too much talk, both which I here confess—
A warning good; hereafter I'll say less.
Let's now be friends; it's time our spite were spent,
Lest we too late this rashness do repent.
Such premises will force a sad conclusion;
Unless we agree, all falls into confusion.
Let Sanguine with her hot hand Choler hold;
To take her moist my moisture will be bold;

My cold cold Melancholy's hand shall clasp;
Her dry dry Choler's other hand shall grasp:
Two hot, two moist, two cold, two dry, here be,
A golden ring, the posey Unity.
Or jars or scoffs let none hereafter see,
But all admire our perfect amity;
Nor be discerned, here's Water, Earth, Air, Fire,
But here so compact body, whole, entire."

This loving counsel pleased them all so well
That Phlegm was judged, for kindness, to excel.

OF THE FOUR AGES OF MAN.

Lo, now four other act upon the stage:
Childhood and Youth, the Manly and Old Age.
The first, son unto Phlegm, grand-child to Water,
Unstable, supple, cold, and moist 's his nature;
The second, Frolic, claims his pedigree
From Blood and Air, for hot and moist is he;
The third of Fire and Choler is composed —
Vindictive he, and quarrelsome disposed;
The last of Earth and heavy Melancholy,
Solid, hating all lightness and all folly.
Childhood was clothed in white and green, to show
His spring was intermixéd with some snow;
Upon his head nature a garland set
Of primrose, daisy, and the violet;
(Such cold, mean flowers the spring puts forth betime
Before the sun hath throughly heat the clime;)
His hobby striding, did not ride, but run;
And in his hand an hour-glass new begun
In danger every moment of a fall —
And when 't is broke then ends his life and all.

The Four Ages

But if he hold till it have run its last,
Then may he live out threescore years or past.
Next Youth came up in gorgeous attire,
As that fond age doth most of all desire,
His suit of crimson, and his scarf of green.
His pride in his countenance was quickly seen;
Garland of roses, pinks, and gillyflowers
Seemed on his head to grow, bedewed with showers;
His face as fresh as is Aurora fair
When, blushing, she first begins to light the air;
No wooden horse, but one of mettle tried,
He seems to fly or swim, and not to ride;
Then, prancing on the stage, about he wheels—
But as he went death waited at his heels.
The next came up in a much graver sort,
As one that caréd for a good report;
His sword by his side, and choler in his eyes,
But neither used as yet, for he was wise;
Of autumn's fruits a basket on his arm,
His golden god in his purse, which was his charm.
And last of all to act upon this stage,
Leaning upon his staff, came up Old Age.
Under his arm a sheaf of wheat he bore,
An harvest of the best. What needs he more?
In his other hand a glass e'en almost run,
Thus writ about,*" This out, then am I done."*
His hoary hairs and grave aspect made way,
And all gave ear to what he had to say.

These being met, each in his equipage,
Intend to speak according to their age;
But wise Old Age did with all gravity
To childish Childhood give precedency,
And to the rest his reason mildly told —
That he was young before he grew so old.
To do as he each one full soon assents.
Their method was that of the Elements —
That each should tell what of himself he knew,
Both good and bad, but yet no more than 's true.
With heed now stood three ages of frail man
To hear the child, who, crying, thus began.

CHILDHOOD.

"Ah, me! conceived in sin, and born with sorrow,
A nothing, here to-day and gone to-morrow,
Whose mean beginning blushing can't reveal,
But night and darkness must with shame conceal!
My mother's breeding sickness I will spare,
Her nine months' weary burden not declare;
To show her bearing pains I should do wrong,
To tell those pangs which can't be told by tongue.
With tears into the world I did arrive;
My mother still did waste as I did thrive,
Who yet, with love and all alacrity,
Spending, was willing to be spent for me.
With wayward cries I did disturb her rest,
Who sought still to appease me with the breast;

With weary limbs she danced and "By, By," sung,
When wretched I, ingrate, had done the wrong.
When infancy was past, my childishness
Did act all folly that it could express;
My silliness did only take delight
In that which riper age did scorn and slight—
In rattles, baubles, and such toyish stuff.
My then ambitious thoughts were low enough;
My high-born soul so straitly was confined
That its own worth it did not know nor mind;
This little house of flesh did spacious count,
Through ignorance all troubles did surmount.
Yet this advantage had mine ignorance—
Freedom from envy and from arrogance.
How to be rich or great I did not cark;
A baron or a duke ne'er made my mark,
Nor studious was kings' favors how to buy
With costly presents or base flattery;
No office coveted, wherein I might
Make strong myself and turn aside weak right;
No malice bore to this or that great peer,
Nor unto buzzing whisperers gave ear.
I gave no hand nor vote for death or life;
I 'd naught to do 'twixt king and people's strife;
No statist I, nor martialist in the field;
Where'er I went, mine innocence was shield.
My quarrels not for diadems did rise,
But for an apple, plum, or some such prize;

My strokes did cause no blood, no wounds or scars;
My little wrath did end soon as my wars;
My duel was no challenge, nor did seek
My foe should weltering in his bowels reek.
I had no suits at law neighbors to vex,
Nor evidence for lands did me perplex.
I feared no storms, nor all the wind that blows;
I had no ships at sea, nor freights to lose.
I feared no drought nor wet—I had no crop;
Nor yet on future things did set my hope.
This was mine innocence. But, ah! the seeds
Lay rakéd up of all the cursed weeds,
Which sprouted forth in mine ensuing age,
As he can tell that next comes on the stage.
But yet let me relate, before I go,
The sins and dangers I am subject to.
Stainéd from birth with Adam's sinful fact,
Thence I began to sin as soon as act:
A perverse will, a love to what 's forbid,
A serpent's sting in pleasing face lay hid;
A lying tongue as soon as it could speak,
And fifth commandment I do daily break;
Oft stubborn, peevish, sullen, pout and cry,
Then naught can please, and yet I know not why.
As many are my sins, so dangers, too;
For sin brings sorrow, sickness, death, and woe;
And though I miss the tossings of the mind,
Yet griefs in my frail flesh I still do find.

What gripes of wind mine infancy did pain!
What tortures I in breeding teeth sustain!
What crudities my stomach cold hath bred,
Whence vomits, flux, and worms have issuéd!
What breaches, knocks, and falls I daily have,
And some perhaps I'll carry to my grave;
Sometimes in fire, sometimes in water, fall,
Strangely preserved, yet mind it not at all.
At home, abroad, my dangers manifold
That wonder 't is my glass till now doth hold.
I've done; unto my elders I give way;
For 't is but little that a child can say.

YOUTH.

My goodly clothing and my beauteous skin
Declare some greater riches are within.
But what is best I'll first present to view,
And then the worst in a more ugly hue.
For thus to do we on this stage assemble;
Then let not him that hath most craft dissemble.
My education and my learning such
As might myself and others profit much:
With nurture trainéd up in virtue's schools,
Of science, arts, and tongues I know the rules;
The manners of the court I also know,
And so likewise what they in the country do.
The brave attempts of valiant knights I prize
That dare scale walls and forts reared to the skies;

The snorting horse, the trumpet, drum, I like,
The glittering sword, the pistol, and the pike.
I cannot lie intrenched before a town,
Nor wait till good success our hopes doth crown.
I scorn the heavy corselet, musket-proof;
I fly to catch the bullet that's aloof.
Though thus in field, at home, to all most kind,
So affable that I can suit each mind,
I can insinuate into the breast,
And by my mirth can raise the heart depressed.
Sweet music wraps my brave, harmonious soul;
My high thoughts elevate beyond the pole;
My wit, my bounty, and my courtesy
Make all to place their future hopes on me.
This is my best; but youth is known, alas,
To be as wild as is the snuffing ass,
As vain as froth or vanity can be,
That who would see vain man may look on me:
My gifts abused, my education lost,
My woeful parents' longing hopes are crossed;
My wit evaporates in merriment;
My valor in some beastly quarrel's spent.
My lust doth hurry me to all that's ill.
I know no law nor reason but my will.—
Sometimes lay wait to take a wealthy purse,
Or stab the man in his own defence—that's worse.
Sometimes I cheat, unkind, a female heir
Of all at once, who, not so wise as fair,

Trusteth my loving looks and glozing tongue
Until her friends, treasure, honor, are gone.
Sometimes I sit carousing others' health
Until mine own be gone, my wit, and wealth;
From pipe to pot, from pot to words and blows,—
For he that loveth wine wanteth no woes,—
Whole nights with ruffians, roarers, fiddlers, spend.
To all obscenity mine ears I lend;
All counsel hate which tends to make me wise,
And dearest friends count for mine enemies.
If any care I take, 't is to be fine;
For sure my suits more than my virtues shine.
If time from lewd companions I can spare,
'T is spent to curl and pounce my new-bought hair.
Some new Adonis I do strive to be;
Sardanapalus now survives in me.
Cards, dice, and oaths concomitant I love;
To plays, to masks, to taverns, still I move.
And in a word, if what I am you 'd hear,
Seek out a British brutish cavalier:
Such wretch, such monster, am I; but yet more—
I have no heart at all this to deplore;
Remembering not the dreadful day of doom,
Nor yet that heavy reckoning soon to come,
Though dangers do attend me every hour,
And ghastly death oft threats me with his power:
Sometimes by wounds in idle combats taken,
Sometimes with agues all my body shaken;

Sometimes by fevers all my moisture drinking,
My heart lies frying, and mine eyes are sinking.
Sometimes the quinsy, painful pleurisy,
With sad affrights of death doth menace me;
Sometimes the two-fold pox me sore bemars
With outward marks and inward loathsome scars;
Sometimes the frenzy strangely mads my brain,
That oft for it in bedlam I remain.
Too many my diseases to recite.
The wonder is I yet behold the light;
That yet my bed in darkness is not made,
And I in black oblivion's den now laid —
Of aches my bones are full, of woe my heart
Clapped in that prison never thence to start.
Thus I have said; and what I 've been, you see.
Childhood and Youth are vain, yea, vanity.

MIDDLE AGE.

Childhood and Youth, forgot, I 've sometime seen,
And now am grown more staid who have been green.
What they have done, the same was done by me;
As was their praise or shame, so mine must be.
Now age is more, more good you may expect;
But more mine age, the more is my defect.
When my wild oats were sown, and ripe, and mown,
I then received an harvest of mine own.
My reason then bade judge how little hope
My empty seed should yield a better crop.

The Four Ages

Then with both hands I grasped the world together,
Thus out of one extreme into another;
But yet laid hold on virtue seemingly —
Who climbs without hold, climbs dangerously.
Be my condition mean, I then take pains
My family to keep, but not for gains.
A father I, for children must provide;
But if none, then for kindred near allied.
If rich, I'm urgéd then to gather more
To bear a part in the world, and feed the poor.
If noble, then mine honor to maintain;
If not, riches nobility can gain.
For time, for place, likewise for each relation,
I wanted not my ready allegiance.
Yet all my powers for self ends are not spent;
For hundreds bless me for my bounty lent
Whose backs I've clothed and bellies I have fed
With mine own fleece and with my household bread.
Yea, justice have I done; was I in place
To cheer the good, and wicked to deface.
The proud I crushed, the oppresséd I set free,
The liars curbed, but nourished verity.
Was I a pastor, I my flock did feed,
And gently led the lambs as they had need.
A captain I, with skill I trained my band,
And showed them how in face of foes to stand.
A soldier I, with speed I did obey
As readily as could my leader say.

Was I a laborer, I wrought all day
As cheerfully as e'er I took my pay.
Thus hath mine age in all sometimes done well;
Sometimes again mine age been worse than hell —
In meanness, greatness, riches, poverty,
Did toil, did broil, oppressed, did steal and lie.
Was I as poor as poverty could be,
Then baseness was companion unto me,
Such scum as hedges and highways do yield,
As neither sow, nor reap, nor plant, nor build.
If to agriculture I was ordained,
Great labors, sorrows, crosses, I sustained.
The early cock did summon but in vain
My wakeful thoughts up to my painful gain.
My weary beast rest from his toil can find;
But if I rest the more distressed my mind.
If happiness my sordidness hath found,
'T was in the crop of my manuréd ground,
My thriving cattle and my new milch cow,
My fleecéd sheep, and fruitful farrowing sow.
To greater things I never did aspire;
My dunghill thoughts or hopes could reach no higher.
If to be rich or great it was my fate,
How was I broiled with envy and with hate!
Greater than was the greatest, was my desire,
And thirst for honor set my heart on fire;
And by ambition's sails I was so carried
That over flats, and sands, and rocks I hurried —

Oppressed, and sunk, and staved all in my way
That did oppose me to my longéd bay.
My thirst was higher than nobility—
I oft longed sore to taste of royalty.
Then kings must be deposed or put to flight,
I might possess that throne which was their right;
There set, I rid myself straight out of hand
Of such competitors as might in time withstand,
Then thought my state firm founded, sure to last.
But in a trice 't is ruined by a blast;
Though cemented with more than noble blood,
The bottom naught, and so no longer stood.
Sometimes vain glory is the only bait
Whereby my empty soul is lured and caught.
Be I of wit, of learning, and of parts,
I judge I should have room in all men's hearts.
And envy gnaws if any do surmount:
I hate not to be held in highest account.
If, Bias like, I 'm stripped unto my skin,
I glory in my wealth I have within.
Thus good and bad, and what I am, you see.
Now, in a word, what my diseases be:
The vexing stone in bladder and in reins;
The strangury torments me with sore pains;
The windy colic oft my bowels rend
To break the darksome prison where it 's penned;
The cramp and gout doth sadly torture me,
And the restraining lame sciatica;

The asthma, megrim, palsy, lethargy,
The quartan ague, dropsy, lunacy:
Subject to all distempers, that's the truth,
Though some more incident to Age or Youth.
And to conclude I may not tedious be:
Man at his best estate is vanity.

OLD AGE.

What you have been, e'en such have I before;
And all you say, say I, and somewhat more.
Babe's innocence, Youth's wildness, I have seen,
And in perplexéd Middle Age have been;
Sickness, dangers, anxieties, have passed,
And on this stage am come to act my last.
I have been young, and strong, and wise as you,
But now *Bis pueri senes* is too true.
In every age I've found much vanity;
An end of all perfection now I see.
It's not my valor, honor, nor my gold
My ruined house, now falling, can uphold;
It's not my learning, rhetoric, wit so large
Hath now the power death's warfare to discharge;
It's not my goodly state, nor bed of down,
That can refresh or ease if conscience frown;
Nor from alliance can I now have hope.
But what I have done well, that is my prop.
He that in youth is godly, wise, and sage
Provides a staff then to support his age.

The Four Ages

Mutations great, some joyful and some sad,
In this short pilgrimage I oft have had.
Sometimes the heavens with plenty smiled on me;
Sometimes, again, rained all adversity;
Sometimes in honor, sometimes in disgrace;
Sometimes an abject, then again in place.
Such private changes oft mine eyes have seen.
In various times of state I 've also been:
I 've seen a kingdom flourish like a tree,
When it was ruled by that celestial she,
And, like a cedar, others so surmount
That but for shrubs they did themselves account.
Then saw I France and Holland saved, Calais won,
And Philip and Albertus half undone.
I saw all peace at home, terror to foes.
But, ah! I saw at last those eyes to close;
And then methought the day at noon grew dark
When it had lost that radiant sun-like spark.
In midst of griefs I saw our hopes revive,
For 't was our hopes then kept our hearts alive.
We changed our queen for king, under whose rays
We joyed in many blest and prosperous days.
I 've seen a prince, the glory of our land,
In prime of youth seized by heaven's angry hand,
Which filled our hearts with fears, with tears our eyes,
Wailing his fate and our own destinies.
I 've seen from Rome an execrable thing —
A plot to blow up nobles and their king;

But saw their horrid faƈt soon disappointed,
And land and nobles saved, with their anointed.
I 've princes seen to live on others' lands;
A royal one by gifts from strangers' hands
Admired for their magnanimity,
Who lost a princedom and a monarchy.
I 've seen designs for Ré and Rochelle crossed,
And poor Palatinate forever lost.
I 've seen unworthy men advancéd high,
And better ones suffer extremity;
But neither favor, riches, title, state,
Could length their days or once reverse their fate —
I 've seen one stabbed, and some to lose their heads,
And others flee, struck both with guilt and dread.
I 've seen, and so have you, for 't is but late,
The desolation of a goodly state,
Plotted and aƈted so that none can tell
Who gave the counsel but the prince of hell —
Three hundred thousand slaughtered innocents
By bloody popish, hellish miscreants.
Oh, may you live, and so you will, I trust,
To see them swill in blood until they burst.
I 've seen a king by force thrust from his throne,
And an usurper subtilely mount thereon.
I 've seen a state unmolded, rent in twain;
But you may live to see it made up again.
I 've seen it plundered, taxed, and soaked in blood;
But out of evil you may see much good.

The Four Ages

What are my thoughts this is no time to say.
Men may more freely speak another day.
These are no old-wives' tales, but this is truth;
We old men love to tell what was done in youth.
But I return from whence I stepped awry.
My memory is bad, my brain is dry;
Mine almond tree, gray hairs, do flourish now,
And back, once straight, apace begins to bow;
My grinders now are few, my sight doth fail,
My skin is wrinkled, and my cheeks are pale.
No more rejoice at music's pleasing noise,
But, waking, glad to hear the cock's shrill voice.
I cannot scent savors of pleasant meat,
Nor sapors find in what I drink or eat.
My arms and hands, once strong, have lost their might;
I cannot labor, much less can I fight.
My comely legs, once nimble as the roe,
Now stiff and numb can hardly creep or go.
My heart, sometime as fierce as lion bold,
Now trembling is, all fearful, sad, and cold.
My golden bowl and silver cord ere long
Shall both be broke by racking death so strong:
Then shall I go whence I shall come no more,
Sons, nephews, leave my farewell to deplore.
In pleasures and in labors I have found
That earth can give no consolation sound
To great, to rich, to poor, to young, to old,
To mean, to noble, to fearful, or to bold:

From king to beggar, all degrees shall find
But vanity, vexation of the mind.
Yea, knowing much, the pleasantest life of all
Hath yet among those sweets some bitter gall:
Though reading others' works doth much refresh,
Yet studying much brings weariness to the flesh.
My studies, labors, readings, all are done,
And my last period now e'en almost run.
Corruptiön my father I do call,
Mother and sisters both the worms that crawl.
In my dark house such kindred I have store
Where I shall rest till heavens shall be no more.
And when this flesh shall rot and be consumed,
This body by this soul shall be assumed,
And I shall see with these same very eyes
My strong Redeemer coming in the skies.
Triumph I shall o'er sin, o'er death, o'er hell.
And in that hope I bid you all farewell.

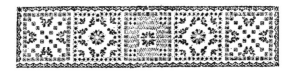

THE FOUR SEASONS OF THE YEAR.

SPRING.

Another four I 've left yet to bring on,
Of four times four the last quaternion,
The Winter, Summer, Autumn, and the Spring;
In season all these seasons I shall bring.
Sweet Spring, like man in his minority,
At present claimed and had priority.
With smiling face, and garments somewhat green,
She trimmed her locks, which late had frosted been;
Nor hot nor cold she spake, but with a breath
Fit to revive the numbéd earth from death.
"Three months," quoth she, "are allotted to my
    share —
March, April, May of all the rest most fair.
Tenth of the first, Sol into Aries enters,
And bids defiance to all tedious winters;
Crosses the line, and equals night and day;
Still adds to the last till after pleasant May;
And now makes glad the darkened northern wights
Who for some months have seen but starry lights.

Now goes the plowman to his merry toil
He might unloose his winter-lockéd soil.
The seedsman, too, doth lavish out his grain
In hope the more he casts the more to gain.
The gardener now superfluous branches lops,
And poles erects for his young clambering hops;
Now digs, then sows his herbs, his flowers, and roots,
And carefully manures his trees of fruits.
The Pleiades their influence now give,
And all that seemed as dead afresh do live:
The croaking frogs, whom nipping winter killed,
Like birds now chirp and hop about the field.
The nightingale, the blackbird, and the thrush
Now tune their lays on sprays of every bush.
The wanton frisking kids and soft-fleeced lambs
Do jump and play before their feeding dams;
The tender tops of budding grass they crop;
They joy in what they have, but more in hope —
For though the frost hath lost his binding power,
Yet many a fleece of snow and stormy shower
Doth darken Sol's bright eye, makes us remember
The pinching north-west wind of cold December.
My second month is April, green and fair,
Of longer days, and a more temperate air.
The sun in Taurus keeps his residence,
And with his warmer beams glanceth from thence.
This is the month whose fruitful showers produces
All set and sown for all delights and uses.

The Four Seasons

The pear, the plum, and apple-tree now flourish;
The grass grows long the hungry beast to nourish.
The primrose pale and azure violet
Among the verdurous grass hath nature set,
That when the sun on his love—the earth—doth shine
These might as lace set out her garment fine.
The fearful bird his little house now builds
In trees and walls, in cities and in fields;
The outside strong, the inside warm and neat —
A natural artificer complete.
The clucking hen her chirping chickens leads;
With wings and beak defends them from the gleads.
My next and last is fruitful pleasant May,
Wherein the earth is clad in rich array.
The sun now enters loving Gemini,
And heats us with the glances of his eye;
Our thicker raiment makes us lay aside,
Lest by his fervor we be torrefied.
All flowers the sun now with his beams discloses
Except the double pinks and matchless roses.
Now swarms the busy, witty honey-bee,
Whose praise deserves a page from more than me.
The cleanly housewife's dairy's now in the prime,
Her shelves and firkins filled for winter-time.
The meads with cowslips, honeysuckles, dight,
One hangs his head, the other stands upright;
But both rejoice at heaven's clear smiling face,
More at her showers, which water them a space.

For fruits my season yields the early cherry,
The hasty pea, and wholesome cool strawberry.
More solid fruits require a longer time;
Each season hath his fruit, so hath each clime —
Each man his own peculiar excellence,
But none in all that hath preëminence."

Sweet fragrant Spring, with thy short pittance fly.
Let some describe thee better than can I.
Yet above all this privilege is thine,
Thy days still lengthen without least decline.

SUMMER.

When Spring had done, then Summer did begin,
With melted, tawny face and garments thin,
Resembling Fire, Choler, and Middle Age,
As Spring did Air, Blood, Youth in his equipage.
Wiping the sweat from off her face that ran,
With hair all wet, puffing, she thus began:
"Bright June, July, and August hot are mine.
In the first Sol doth in crabbed Cancer shine;
His progress to the north now's fully done;
Then retrograde must be my burning sun,
Who to his southward tropic still is bent,
Yet doth his parching heat but more augment
Though he decline, because his flames so fair
Have throughly dried the earth and heat the air.

Like as an oven that long time hath been heat,
Whose vehemency at length doth grow so great
That if you do withdraw her burning store
'T is for a time as fervent as before.
Now go those frolic swains, the shepherd lads,
To wash their thick-clothed flocks, with pipes full glad.
In the cool streams they labor with delight,
Rubbing the dirty coats till they look white,
Whose fleece when finely spun and deeply dyed
With robes thereof kings have been dignified.
Blest rustic swains, your pleasant, quiet life
Hath envy bred in kings that were at strife;
Careless of worldly wealth you sing and pipe,
While they 're embroiled in wars and troubles rife;
Which made great Bajazet cry out in his woes,
'O happy shepherd, which hath not to lose
Orthobulus, nor yet Sebastia great,
But whistleth to thy flock in cold and heat;
Viewing the sun by day, the moon by night,
Endymion's, Diana's, dear delight;
Upon the grass resting your healthy limb,
By purling brooks looking how fishes swim.
If pride within your lowly cells e'er haunt,
Of Him that was Shepherd, then King, go vaunt.'
This month the roses are distilled in glasses,
Whose fragrant smell all made perfumes surpasses.
The cherry, gooseberry, now are in their prime,
And for all sorts of pease this is the time.

July's my next, the hottest in all the year.
The sun through Leo now takes his career,
Whose flaming breath doth melt us from afar,
Increaséd by the star canicular.
This month from Julius Cæsar took its name,
By Romans celebrated to his fame.
Now go the mowers to their slashing toil,
The meadows of their riches to despoil;
With weary strokes they take all in their way,
Bearing the burning heat of the long day.
The forks and rakes do follow them amain,
Which makes the aged fields look young again.
The groaning carts do bear away this prize
To stacks and barns, where it for fodder lies.
My next and last is August, fiery hot,
For much the southward sun abateth not.
This month he keeps with Virgo for a space;
The driéd earth is parchéd with his face.
August of great Augustus took its name,
Rome's second emperor, of lasting fame.
With sickles now the bending reapers go
The ruffling tress of terra down to mow,
And bundle up in sheaves the weighty wheat
Which, after, manchet makes for kings to eat.—
The barley, rye, and pease should first had place,
Although their bread have not so white a face.
The carter leads all home with whistling voice;
He plowed with pain, but, reaping, doth rejoice;

The Four Seasons

His sweat, his toil, his careful, wakeful nights,
His fruitful crop abundantly requites.
Now 's ripe the pear, pear-plum, and apricock,
The prince of plums, whose stone 's as hard as rock."

The Summer seems but short; the Autumn hastes
To shake his fruits, of most delicious tastes,
Like good Old Age, whose younger juicy roots
Hath still ascended to bear goodly fruits
Until his head be gray and strength be gone.
Yet then appear the worthy deeds he hath done:
To feed his boughs exhausted he his sap,
Then dropped his fruits into the eater's lap.

AUTUMN.

" Of Autumn's months September is the prime;
Now day and night are equal in each clime.
The twelfth of this Sol riseth in the line,
And doth in poising Libra this month shine.
The vintage now is ripe; the grapes are pressed,
Whose lively liquor oft is cursed and blessed.
For naught so good but it may be abused;
But it 's a precious juice when well it 's used.
The raisins now in clusters dried be,
The orange, lemon, dangle on the tree;
The pomegranate, the fig, are ripe also,
And apples now their yellow sides do show.

Of almonds, quinces, wardens, and the peach
The season's now at hand of all and each.
Sure at this time time first of all began,
And in this month was made apostate man;
For then in Eden was not only seen
Boughs full of leaves, or fruits unripe or green,
Or withered stocks which were all dry and dead,
But trees with goodly fruits replenishéd;
Which shows nor Summer, Winter, nor the Spring
Our grandsire was of Paradise made king.
Nor could that temperate clime such difference
    make,
If sited as the most judicious take.
October is my next. We hear in this
The northern winter-blasts begin to hiss.
In Scorpio resideth now the sun,
And his declining heat is almost done.
The fruitless trees all withered now do stand,
Whose sapless yellow leaves by winds are fanned,
Which notes when youth and strength have passed
    their prime
Decrepit age must also have its time.
The sap doth slyly creep towards the earth,
There rests until the sun give it a birth.
So doth Old Age still tend unto his grave,
Where also he his winter-time must have;
But when the Sun of Righteousness draws nigh,
His dead old stock shall mount again on high.

November is my last, for time doth haste.
We now of Winter's sharpness 'gin to taste.
This month the sun's in Sagittarius,
So far remote his glances warm not us.
Almost at shortest is the shortened day.
The Northern Pole beholdeth not one ray.
Now Greenland (Grönland), Finland, Lapland, see
No sun to lighten their obscurity —
Poor wretches that in total darkness lie,
With minds more dark than is the darkened sky.
Beef, brawn, and pork are now in great request,
And solid meats our stomachs can digest.
This time warm clothes, full diet, and good fire
Our pinchéd flesh and hungry maws require.
Old, cold, dry Age and Earth Autumn resembles,
And Melancholy, which most of all dissembles.
I must be short, and short's the shortened day.
What Winter hath to tell, now let him say."

WINTER.

Cold, moist, young phlegmy Winter now doth lie
In swaddling clouts, like new-born infancy;
Bound up with frosts, and furred with hail and snows,
And, like an infant, still he taller grows.
" December is my first, and now the sun
To the southward tropic his swift race doth run.
This month he's housed in hornéd Capricorn;
From thence begins to length the shortened morn.

Through Christendom with great festivity
Now 's held a guessed but blest Nativity.
Cold, frozen January next comes in,
Chilling the blood and shrinking up the skin.
In Aquarius now keeps the long wished sun,
And northward his unwearied course doth run;
The day much longer than it was before,
The cold not lessened, but augmented more.
Now toes, and ears, and fingers often freeze,
And travelers their noses sometimes leese.
Moist, snowy February is my last.
I care not how the winter-time doth haste.
In Pisces now the golden sun doth shine,
And northward still approaches to the line.
The rivers begin to ope, the snows to melt,
And some warm glances from his face are felt;
Which is increaséd by the lengthened day,
Till by his heat he drive all cold away,"

And thus the year in circle runneth round.
Where first it did begin, in the end it 's found.
My subject 's bare, my brain is bad,
Or better lines you should have had.
The first fell in so naturally,
I knew not how to pass it by;
The last, though bad, I could not mend.
Accept, therefore, of what is penned,
And all the faults that you shall spy
Shall at your feet for pardon cry.

THE FOUR MONARCHIES.
THE ASSYRIAN BEING THE FIRST, BEGINNING UNDER NIMROD, ONE HUNDRED AND THIRTY-ONE YEARS AFTER THE FLOOD.

When time was young, and the world in infancy,
Man did not proudly strive for sovereignty;
But each one thought his petty rule was high
If of his house he held the monarchy.
This was the Golden Age; but after came
The boisterous son of Cush, grandchild to Ham,
That mighty hunter who in his strong toils
Both beasts and men subjected to his spoils,
The strong foundation of proud Babel laid,
Erech, Accad, and Calneh also made.
These were his first; all stood in Shinar land.
From thence he went Assyria to command,
And mighty Nineveh he there begun,
Not finishéd till he his race had run;
Resen, Calah, and Rehoboth, likewise,
By him to cities eminent did rise.

Of Saturn he was the original,
Whom the succeeding times a god did call.
When thus with rule he had been dignified,
One hundred fourteen years he after died.

BELUS.

Great Nimrod dead, Belus the next, his son,
Confirms the rule his father had begun;
Whose acts and power are not for certainty
Left to the world by any history.
But yet this blot for ever on him lies —
He taught the people first to idolize.
Titles divine he to himself did take.
Alive and dead a god they did him make.
This is that Bel the Chaldees worshipéd,
Whose priests in stories oft are mentionéd;
This is that Baal to whom the Israelites
So oft profanely offered sacred rites;
This is Beelzebub, god of Ekronites;
Likewise Baalpeor, of the Moabites.
His reign was short, for, as I calculate,
At twenty-five ended his regal date.

NINUS.

His father dead, Ninus begins his reign,
Transfers his seat to the Assyrian plain,
And mighty Nineveh more mighty made
Whose foundation was by his grandsire laid:

Four hundred forty furlongs walled about,
On which stood fifteen hundred towers stout;
The walls one hundred sixty feet upright,
So broad three chariots run abreast there might.
Upon the pleasant banks of Tigris' flood
This stately seat of warlike Ninus stood.
This Ninus for a god his father canonized,
To whom the sottish people sacrificed.
This tyrant did his neighbors all oppress;
Where'er he warred he had too good success —
Barzanes, the great Armenian king,
By force and fraud did under tribute bring;
The Median country he did also gain,
Pharnus, their king, he caused to be slain;
An army of three millions he led out
Against the Bactrians (but that I doubt);
Zoroaster, their king, he likewise slew,
And all the greater Asia did subdue.
Semiramis from Menon did he take;
Then drowned himself did Menon for her sake.
Fifty-two years he reigned, as we are told.
The world then was two thousand nineteen old.

SEMIRAMIS.

This great oppressing Ninus dead and gone,
His wife Semiramis usurped the throne;
She like a brave virago played the rex,
And was both shame and glory of her sex.

Her birthplace was Philistia's Ascalon,
Her mother Derceto, a courtezan.—
Others report she was a vestal nun
Adjudgéd to be drowned for the crime she'd done,
Transformed into a fish by Venus' will,
Her beauteous face, they feign, retaining still.
Sure from this fiction Dagon first began,
Changing the woman's face into a man.
But all agree that from no lawful bed
This great renownéd empress issuéd;
For which she was obscurely nourishéd —
Whence rose that fable she by birds was fed.
This gallant dame unto the Bactrian war
Accompanying her husband Menon far,
Taking a town such valor she did show
That Ninus amorous of her soon did grow,
And thought her fit to make a monarch's wife,
Which was the cause poor Menon lost his life.
She flourishing with Ninus long did reign,
Till her ambition caused him to be slain
That, having no compeer, she might rule all,
Or else she sought revenge for Menon's fall.
Some think the Greeks this slander on her cast,
As on her life licentious and unchaste;
That, undeserved, they blurred her name and fame
By their aspersions cast upon the same.
But were her virtues more or less or none,
She for her potency must go alone.

Her wealth she showed in building Babylon,
Admired of all, but equalized of none;
The wall so strong and curiously was wrought,
That after ages skill by it was taught.
With towers and bulwarks made of costly stone,
Quadrangle was the form it stood upon.
Each square was fifteen thousand paces long.
An hundred gates it had of metal strong.
Three hundred sixty feet the wall in height,
Almost incredible it was in breadth —
Some writers say six chariots might a-front
With great facility march safe upon 't.
About the wall a ditch so deep and wide
That like a river long it did abide;
Three hundred thousand men here day by day
Bestowed their labor and received their pay.
And that which did all cost and art excel
The wondrous temple was she reared to Bel,
Which in the midst of this brave town was placed,
Continuing till Xerxes it defaced;
Whose stately top above the clouds did rise,
From whence astrologers oft viewed the skies.
This to describe in each particular,
A structure rare I should but rudely mar.
Her gardens, bridges, arches, mounts, and spires
Each eye that saw or ear that heard admires.
In Shinar plain, on the Euphratean flood,
This wonder of the world, this Babel, stood.

An expedition to the east she made
Stratobatis his country to invade.
Her army of four millions did consist
(Each may believe it as his fancy list);
Her camels, chariots, galleys, in such number
As puzzles best historians to remember.
But this is wonderful — of all those men
They say but twenty e'er came back again;
The river Indus swept them half away,
The rest Stratobatis in fight did slay.
This was last progress of this mighty queen,
Who in her country never more was seen.
The poets feigned her turned into a dove,
Leaving the world to Venus, soared above;
Which made the Assyrians many a day
A dove within their ensigns to display.
Forty-two years she reigned, and then she died,
But by what means we are not certified.

NINIAS, OR ZAMES.

His mother dead, Ninias obtains his right,
A prince wedded to ease and to delight,
Or else was his obedience very great
To sit thus long, obscure, robbed of his seat.
Some write his mother put his habit on,
Which made the people think they served her son;
But much it is, in more than forty years
This fraud in war nor peace at all appears.

More like it is, his lust with pleasures fed,
He sought no rule till she was gone and dead.
What then he did of worth can no man tell,
But is supposed to be that Amraphel
Who warred with Sodom's and Gomorrah's king,
'Gainst whom his trainéd bands Abram did bring.
But this is far unlike, he, being son
Unto a father that all countries won,
So suddenly should lose so great a state
With petty kings to join confederate.
Nor can those reasons which wise Raleigh finds
Well satisfy the most considerate minds.
We may with learned Usher better say
He many ages lived after that day;
And that Semiramis then flourishéd
When famous Troy was so beleaguéred.
Whate'er he was, or did, or how it fell,
We may suggest our thoughts, but cannot tell;
For Ninias and all his race are left
In deep oblivion, of acts bereft,
And many hundred years in silence sit,
Save a few names a new Berosus writ.
And such as care not what befalls their fames
May feign as many acts as he did names.
It may suffice, if all be true that's past.
To Sardanapalus next we will make haste.

SARDANAPALUS.

Sardanapalus, son to Ocrazapes,
Who wallowéd in all voluptuousness,—
That palliardizing sot that out of doors
Ne'er showed his face, but reveled with his whores,
Did wear their garbs, their gestures imitate,
And in their kind to excel did emulate,—
His baseness knowing, and the people's hate,
Kept close, fearing his well deservéd fate.
It chanced Arbaces brave unwarily
His master like a strumpet clad did spy;
His manly heart disdainéd in the least
Longer to serve this metamorphosed beast.
Unto Belesis then he brake his mind,
Whom sick of his disease he soon did find.
These two ruled Media and Babylon;
Both for their king held their dominiön.
Belesis promiséd Arbaces aid,
Arbaces him fully to be repaid.
The last the Medes and Persians does invite
Against their monstrous king to use their might.
Belesis the Chaldeans doth require
And the Arabians, to further his desire.
These all agree, and forty thousand make,
The rule from their unworthy prince to take.
These forces musteréd, and in array,
Sardanapalus leaves his apish play;

And though of wars he did abhor the sight,
Fear of his diadem did force him fight;
And either by his valor or his fate
Arbaces' courage he did so abate
That in despair he left the field and fled,
But with fresh hopes Belesis succoréd.
From Bactria an army was at hand
Pressed for this service by the king's command.
These with celerity Arbaces meets,
And with all terms of amity them greets;
With promises their necks now to unyoke,
And their taxations sore all to revoke;
To enfranchise them, to grant what they could crave,
No privilege to want, subjects should have —
Only entreats them to join their force with his,
And win the crown, which was the way to bliss.
Won by his loving looks, more by his speech,
To accept of what they could they all beseech.
Both sides their hearts, their hands, and bands unite,
And set upon their prince's camp that night;
Who, reveling in cups, sung care away
For victory obtained the other day,
And now, surprised by this unlooked-for fright,
Bereft of wits, are slaughteréd downright.
The king his brother leaves all to sustain,
And speeds himself to Nineveh amain.
But, Salmoneus slain, the army falls;
The king's pursued unto the city's walls.

But he once in, pursuers came too late,
The walls and gates their haste did terminate.
There with all store he was so well provided
That what Arbaces did was but derided,
Who there encamped two years for little end.
But in the third the river proved his friend;
For by the rain was Tigris so o'erflown
Part of that stately wall was overthrown.
Arbaces marches in, the town he takes,
For few or none, it seems, resistance makes.
And now they saw fulfilled a prophecy,
That when the river proved their enemy
Their strong walled town should suddenly be taken.
By this accomplishment their hearts were shaken.
Sardanapalus did not seek to fly
This his inevitable destiny;
But all his wealth and friends together gets,
Then on himself and them a fire he sets.
This was last monarch of great Ninus' race,
That for twelve hundred years had held the place;
Twenty he reigned, same time, as stories tell,
That Amaziah was king of Israel.
His father was then king, as we suppose,
When Jonah for their sins denounced those woes;
He did repent, the threatening was not done,
But now accomplished in his wicked son.
Arbaces, thus of all becoming lord,
Ingenuously with all did keep his word.

Of Babylon Belesis he made king,
With overplus of all the wealth therein.
To Bactrians he gave their liberty.
Of Ninevites he causéd none to die,
But suffered with their goods to go elsewhere,
Not granting them now to inhabit there;
For he demolishéd that city great,
And unto Media transferred his seat.
Such was his promise which he firmly made
To Medes and Persians when he craved their aid.
A while he and his race aside must stand,
Not pertinent to what we have in hand;
And Belochus in his progeny pursue,
Who did this monarchy begin anew.

BELESIS, OR BELOCHUS.

Belesis settled in his new old seat
Not so content, but aiming to be great,
Encroaching still upon the bordering lands
Till Mesopotamia he got in his hands.
And, either by compound or else by strength,
Assyria he gained also at length;
Then did rebuild destroyéd Nineveh,
A costly work which none could do but he
Who owned the treasures of proud Babylon
And those that seemed with Sardanapalus gone.
For though his palace did in ashes lie,
The fire those metals could not damnify;

From rubbish these with diligence he rakes.
Arbaces suffers all, and all he takes.
He, thus enriched by this new-triéd gold,
Raises a phenix new from grave of the old;
And from this heap did after ages see
As fair a town as the first Nineveh.
When this was built, and matters all in peace,
Molests poor Israel, his wealth to increase:
A thousand talents of Menahem had,
Who to be rid of such a guest was glad.
In sacred writ he's known by name of Pul,
Which makes the world of difference so full
That he and Belochus could not one be;
But circumstance doth prove the verity,
And times of both computed so fall out
That these two made but one we need not doubt.
What else he did his empire to advance
To rest content we must in ignorance.
Forty-eight years he reigned, his race then run.
He left his new-got kingdom to his son.

TIGLATH-PILESER.

Belesis dead, Tiglath, his warlike son,
Next treads those steps by which his father won.
Damascus, ancient seat of famous kings,
Under subjection by his sword he brings;
Rezin, their valiant king, he also slew,
And Syria to obedience did subdue.

The Four Monarchies

Judah's bad king occasionéd this war
When Rezin's force his borders sore did mar
And divers cities by strong hand did seize.
To Tiglath then doth Ahaz send for ease;
The temple robs so to fulfil his ends,
And to Assyria's king a present sends.
"I am thy servant and thy son," quoth he;
"From Rezin and from Pekah set me free."
Gladly doth Tiglath this advantage take,
And succors Ahaz, yet for Tiglath's sake.
Then, Rezin slain, his army overthrown,
He Syria makes a province of his own.
Unto Damascus then comes Judah's king
His humble thankfulness in haste to bring,
Acknowledging the Assyrian's high desert,
To whom he ought all loyalty of heart.
But Tiglath, having gained his wishéd end,
Proves unto Ahaz but a feignéd friend;
All Israel's lands beyond Jordan he takes,
In Galilee he woeful havoc makes.
Through Syria now he marched; none stopped his way,
And Ahaz open at his mercy lay,
Who still implored his love, but was distressed.
This was that Ahaz who so high transgressed.
Thus Tiglath reigned and warred twenty-seven years.
Then by his death released were Israel's fears.

SALMANASSAR, OR NABONASSAR.

Tiglath deceased, Salmanassar was next.
He Israelites more than his father vexed.
Hosea, their last king, he did invade,
And him six years his tributary made.
But, weary of his servitude, he sought
To Egypt's king, which did avail him naught;
For Salmanassar, with a mighty host,
Besieged his regal town, and spoiled his coast,
And did the people, nobles, and their king
Into perpetual thraldom that time bring.
Those that from Joshua's time had been a state
Did justice now by him eradicate;
This was that strange, degenerated brood
On whom nor threats nor mercies could do good.
Laden with honor, prisoners, and spoil,
Returns triumphant victor to his soil;
He placéd Israel there where he thought best,
Then sent his colonies theirs to invest.
Thus Jacob's sons in exile must remain,
And pleasant Canaan never see again.
Where now those ten tribes are can no man tell,
Or how they fare, rich, poor, or ill, or well;
Whether the Indians of the East or West,
Or wild Tartarians, as yet ne'er blest,
Or else those Chinese rare, whose wealth and arts
Have bred more wonder than belief in hearts.

But what or where they are, yet know we this —
They shall return, and Sion see with bliss.

SENNACHERIB.

Sennacherib Salmanassar succeeds,
Whose haughty heart is shown in words and deeds.
His wars none better than himself can boast,
On Hena, Arpad, and on Judah's coast,
On Ivah's and on Sepharvaim's gods;
'Twixt them and Israel's he knew no odds
Until the thundering hand of heaven he felt,
Which made his army into nothing melt.
With shame he turned to Nineveh again,
And by his sons in his idols' house was slain.

ESARHADDON.

His son, weak Esarhaddon, reigned in his place,
The fifth and last of great Belesis' race.
Brave Merodach, the son of Baladan,
In Babylon lieutenant to this man,
Of opportunity advantage takes,
And on his master's ruins his house makes;
As Belesis his sovereign did unthrone,
So he's now styled the king of Babylon.
After twelve years did Esarhaddon die,
And Merodach assumed the monarchy.

MERODACH-BALADAN.

All yield to him but Nineveh, kept free
Until his grandchild made her bow the knee.
Ambassadors to Hezekiah he sent,
His health to congratulate with compliment.

BEN-MERODACH.

Ben-Merodach, successor to this king,
Of him is little said in anything,
But by conjecture this, that none but he
Led King Manasseh to captivity.

NABOPOLASSAR.

Brave Nabopolassar to this king was son.
The famous Nineveh by him was won;
For fifty years, or more, she had been free,
But yields her neck now to captivity.
A viceroy from her foe she's glad to accept,
By whom in firm obedience she's kept.
This king's less famed for all the acts he's done
Than being father to so great a son.

NEBUCHADNEZZAR.

The famous acts of this heroic king
Did neither Homer, Hesiod, Virgil, sing;
Nor of his wars have we the certainty
From some Thucydides' grave history;
Nor his metamorphosis from Ovid's book,
Nor his restoring from old legends took;

The Four Monarchies

But by the prophets, penmen most divine,
This prince in his magnitude doth ever shine.
This was of monarchies that head of gold,
The richest and the dreadfullest to behold;
This was that tree whose branches filled the earth,
Under whose shadow birds and beasts had birth;
This was that king of kings did what he pleased,
Killed, saved, pulled down, set up, or pained, or eased;
And this was he who, when he feared the least,
Was changéd from a king into a beast.
This prince the last years of his father's reign
Against Jehoiakim marched with his train.
Judah's poor king, besieged and succorless,
Yields to his mercy, and the present stress;
His vassal is, gives pledges for his truth,
Children of royal blood, unblemished youth.
Wise Daniel and his fellows, 'mongst the rest,
By the victorious king to Babel are pressed;
The temple of rich ornaments he defaced,
And in his idols' house the vessels placed.
The next year he, with unresisted hand,
Quite vanquished Pharaoh-Necho with his band:
By great Euphrates did his army fall,
Which was the loss of Syria withal.
Then into Egypt Necho did retire,
Which in few years proves the Assyrian's hire.
A mighty army next he doth prepare,
And unto wealthy Tyre in haste repair.

Such was the situation of this place,
As might not him, but all the world, outface,
That in her pride she knew not which to boast—
Whether her wealth or yet her strength was most.
How in all merchandise she did excel
None but the true Ezekiel need to tell;
And for her strength, how hard she was to gain,
Can Babel's tired soldiers tell with pain.
Within an island had this city seat,
Divided from the main by channel great;
Of costly ships and galleys she had store,
And mariners to handle sail and oar.
But the Chaldeans had nor ships nor skill;
Their shoulders must their master's mind fulfil—
Fetched rubbish from the opposite old town,
And in the channel threw each burden down,
Where, after many essays, they made at last
The sea firm land, whereon the army passed
And took the wealthy town. But all the gain
Requited not the loss, the toil, and pain.
Full thirteen years in this strange work he spent
Before he could accomplish his intent,
And, though a victor, home his army leads
With peeléd shoulders and with balded heads.
When in the Tyrian war this king was hot
Jehoiakim his oath had clean forgot;
Thinks this the fittest time to break his bands,
Whilst Babel's king thus deep engagéd stands.

The Four Monarchies

But he whose fortunes all were in the ebb
Had all his hopes like to a spider's web;
For this great king withdraws part of his force,
To Judah marches with a speedy course,
And, unexpected, finds the feeble prince,
Whom he chastised thus for his proud offence:
Fast bound, intends to Babel him to send,
But changed his mind, and caused his life there end,
Then cast him out like to a naked ass.
For this is he for whom none said Alas!
His son he sufferéd three months to reign,
Then from his throne he plucked him down again;
Whom, with his mother, he to Babel led,
And seven and thirty years in prison fed.
His uncle he established in his place,
Who was last king of holy David's race;
But he as perjured as Jehoiakim,
They lost more now than e'er they lost by him.
Seven years he kept his faith, and safe he dwells,
But in the eighth against his prince rebels.
The ninth came Nebuchadnezzar with power,
Besieged his city, temple, Sion's tower,
And after eighteen months he took them all.
The walls so strong, that stood so long, now fall.
The accurséd king by flight could nowise fly
His well-deserved and foretold misery;
But, being caught, to Babel's wrathful king
With children, wives, and nobles all they bring,

Where to the sword all but himself were put,
And with that woeful sight his eyes close shut.
Ah, hapless man, whose darksome contemplation
Was nothing but such ghastly meditation!
In midst of Babel now till death he lies,
Yet as was told ne'er saw it with his eyes.
The temple's burnt, the vessels had away,
The towers and palaces brought to decay;
Where late of harp and lute were heard the noise,
Now Zim and Jim lift up their screeching voice.
All now of worth are captive led with tears,
And sit bewailing Sion seventy years.
With all these conquests Babel's king rests not,
No, not when Moab, Edom, he had got;
Kedar and Hazor, the Arabians, too,
All vassals, at his hands for grace must sue.
A total conquest of rich Egypt makes,
All rule he from the ancient Pharaohs takes;
Who had for sixteen hundred years borne sway
To Babylon's proud king now yield the day.
Then Put and Lud do at his mercy stand;
Where'er he goes he conquers every land.
His sumptuous buildings pass all conceit,
Which wealth and strong ambition made so great.
His image Judah's captives worship not,
Although the furnace be seven times more hot.
His dreams wise Daniel doth expound full well,
And his unhappy change with grief foretell.

Strange melancholy humors on him lay,
Which for seven years his reason took away;
Which from no natural causes did proceed,
But for his pride, so had the heavens decreed.
The time expired, he brute remains no more,
But government resumes as heretofore;
In splendor and in majesty he sits,
Contemplating those times he lost his wits.
And if by words we may guess at the heart,
This king among the righteous had a part.
Forty-four years he reigned, which, being run,
He left his wealth and conquests to his son.

EVIL-MERODACH.

Babel's great monarch now laid in the dust,
His son possesses wealth and rule as just,
And in the first year of his royalty
Easeth Jehoiachin's captivity;
Poor, forlorn prince, who had all state forgot,
In seven and thirty years had seen no jot.
Among the conquered kings that there did lie
Is Judah's king now lifted up on high;
But yet in Babel he must still remain,
And native Canaan never see again.
Unlike his father, Evil-Merodach
Prudence and magnanimity did lack.
Fair Egypt is by his remissness lost,
Arabia, and all the bordering coast.

Wars with the Medes unhappily he waged,
Within which broils rich Crœsus was engaged;
His army routed, and himself there slain,
His kingdom to Belshazzar did remain.

BELSHAZZAR.

Unworthy Belshazzar next wears the crown,
Whose acts profane a sacred pen sets down;
His lust and cruelties in stories find—
A royal state ruled by a brutish mind.
His life so base and dissolute invites
The noble Persian to invade his rights;
Who, with his own and uncle's power, anon
Lays siege to his regal seat, proud Babylon.
The coward king, whose strength lay in his walls,
To banqueting and reveling now falls,
To show his little dread but greater store,
To cheer his friends and scorn his foes the more.
The holy vessels, thither brought long since,
They caroused in; the sacrilegious prince
Did praise his gods of metal, wood, and stone,
Protectors of his crown and Babylon.
But He, above, his doings did deride,
And with a hand soon dashéd all this pride.
The king upon the wall casting his eye
The fingers of a handwriting did spy,
Which horrid sight he fears must needs portend
Destruction to his crown, to his person end.

With quaking knees and heart appalled he cries
For the soothsayers and magicians wise
This language strange to read and to unfold;
With gifts of scarlet robe, and chain of gold,
And highest dignity next to the king
To him that could interpret clear this thing.
But dumb the gazing astrologers stand,
Amazéd at the writing and the hand.
None answers the affrighted king intent,
Who still expects some fearful sad event.
As dead, alive he sits, as one undone.
In comes the queen to cheer her heartless son;
Of Daniel tells, who in his grandsire's days
Was held in more account than now he was.
Daniel in haste is brought before the king,
Who doth not flatter, nor once cloak the thing;
Reminds him of his grandsire's height and fall,
And of his own notorious sins withal —
His drunkenness, and his profaneness high,
His pride and sottish gross idolatry.
The guilty king, with color pale and dead,
Then hears his "Mene" and his "Tekel" read;
And one thing did worthy a king, though late —
Performed his word to him that told his fate.
That night victorious Cyrus took the town,
Who soon did terminate his life and crown.
With him did end the race of Baladan;
And now the Persian Monarchy began.

The end of the Assyrian Monarchy.

THE SECOND MONARCHY, BEING THE PERSIAN, BEGUN UNDER CYRUS; DARIUS, BEING HIS UNCLE AND FATHER-IN-LAW, REIGNED WITH HIM ABOUT TWO YEARS.

Cyrus, Cambyses' son, of Persia king,
Whom lady Mandane did to him bring;
She, daughter unto great Astyages;
He, in descent, the seventh from Arbaces.
Cambyses was of Achæmenes' race,
Who had in Persia the lieutenant's place
When Sardanapalus was overthrown,
And from that time had held it as his own.
Cyrus Darius' daughter took to wife,
And so unite two kingdoms without strife.
Darius unto Mandane was brother,
Adopts her son for his, having no other.
This is of Cyrus the true pedigree,
Whose ancestors were royal in degree.
His mother's dream, and grandsire's cruelty,
His preservation in his misery,
His nourishment afforded by a bitch,
Are fit for such whose ears for fables itch.
He in his younger days an army led
Against great Crœsus, then of Lydia head;
Who, over-curious of war's event,
For information to Apollo went,

And the ambiguous oracle did trust,—
So overthrown by Cyrus, as was just;
Who him pursues to Sardis, takes the town,
Where all that dare resist are slaughtered down.
Disguiséd, Crœsus hoped to escape in the throng,
Who had no might to save himself from wrong;
But as he passed, his son, who was born dumb,
With pressing grief and sorrow overcome
Among the tumult, bloodshed, and the strife,
Broke his long silence, cried, "Spare Crœsus' life!"
Crœsus thus known, it was great Cyrus' doom—
A hard decree—to ashes he consume.
Then on a woodpile set, where all might eye,
He "Solon! Solon! Solon!" thrice did cry.
The reason of those words Cyrus demands,
Who Solon was, to whom he lifts his hands.
Then to the king he makes this true report:
That Solon sometime at his stately court
His treasures, pleasures, pomp, and power did see,
And, viewing all, at all naught moved was he.
When Crœsus, angry, urged him to express
If ever king equaled his happiness,
Quoth he, "That man for happy we commend
Whose happy life attains a happy end."
Cyrus, with pity moved, knowing a king's stand,
Now up and down, as fortune turns her hand,
Weighing the age and greatness of the prince,—
His mother's uncle, stories do evince,—

Gave him his life and took him for a friend,
Did to him still his chief designs commend.
Next war the restless Cyrus thought upon
Was conquest of the stately Babylon,
Now treble-walled, and moated so about
That all the world they need not fear nor doubt.
To drain this ditch he many sluices cut,
But till convenient time their heads kept shut.
That night Belshazzar feasted all his rout
He cut those banks and let the river out,
And to the walls securely marches on,
Not finding a defendant thereupon;
Entering the town, the sottish king he slays,
Upon earth's richest spoils each soldier preys.
Here twenty years' provision good he found.
Forty-five miles this city scarce could round.
This head of kingdoms, Chaldea's excellence,
For owls and satyrs made a residence;
Yet wondrous monuments this stately queen
A thousand years after had to be seen.
Cyrus doth now the Jewish captives free;
An edict made the temple builded be;
He, with his uncle, Daniel sets on high,
And caused his foes in lions' dens to die.
Long after this he 'gainst the Scythians goes,
And 'Tomyris' son and army overthrows;
Which to revenge she hires a mighty power,
And sets on Cyrus in a fatal hour,

There routs his host, himself a prisoner takes,
And at one blow the world's head headless makes—
The which she bathed within a butt of blood,
Using such taunting words as she thought good.
But Xenophon reports he died in his bed
In honor, peace, and wealth, with a gray head,
And in his town of Pasargadæ lies;
Where some long after sought in vain for prize,
But in his tomb was only to be found
Two Scythian bows, a sword, and target round;
And Alexander, coming to the same,
With honors great did celebrate his fame.
Three daughters and two sons he left behind,
Ennobled more by birth than by their mind.
Thirty-two years in all this prince did reign,
But eight whilst Babylon he did retain;
And though his conquests made the earth to groan,
Now quiet lies under one marble stone,
And with an epitaph himself did make
To show how little land he then should take.

CAMBYSES.

Cambyses, no ways like his noble sire,
Yet to enlarge his state had some desire.
His reign with blood and incest first begins,
Then sends to find a law for these his sins.
That kings with sisters match no law they find
But that the Persian king may act his mind.

He wages war, the fifth year of his reign,
'Gainst Egypt's king, who there by him was slain;
And all of royal blood that came to hand
He seizéd first of life and then of land.
But little Narus 'scaped that cruel fate,
Who, grown a man, resumed again his state.
He next to Cyprus sends his bloody host,
Who, landing soon upon that fruitful coast,
Made Evelthon, their king, with bended knee
To hold his own of his free courtesy.
The temple he destroys, not for his zeal,
For he would be professed god of their weal;
Yea, in his pride, he venturéd so far
To spoil the temple of great Jupiter —
But as they marchéd o'er those desert sands
The stormèd dust o'erwhelmed his daring bands.
But scorning thus by Jove to be outbraved,
A second army he had almost graved;
But vain he found to fight with elements,
So left his sacrilegious bold intents.
The Egyptian Apis then he likewise slew,
Laughing to scorn that sottish calvish crew.
If all this heat had been for pious end,
Cambyses to the clouds we might commend;
But he that 'fore the gods himself prefers
Is more profane than gross idolaters.
He after this, upon suspicion vain,
Unjustly caused his brother to be slain;

The Four Monarchies

Praxaspes into Persia then is sent
To act in secret this his lewd intent.
His sister, whom incestuously he wed,
Hearing her harmless brother thus was dead,
His woeful death with tears did so bemoan
That by her husband's charge she caught her own;
She with her fruit at once were both undone
Who would have borne a nephew and a son.
O hellish husband, brother, uncle, sire,
Thy cruelty all ages will admire.
This strange severity he sometimes used
Upon a judge for taking bribes accused:
Flayed him alive, hung up his stuffèd skin
Over his seat, then placed his son therein,
To whom he gave this in remembrance—
Like fault must look for the like recompense.
His cruelty was come unto that height
He spared nor foe, nor friend, nor favorite.
'T would be no pleasure, but a tedious thing,
To tell the facts of this most bloody king;
Fearèd of all, but loved of few or none,
All wished his short reign past before 't was done.
At last two of his officers, he hears,
Had set one Smerdis up, of the same years
And like in feature to his brother dead,
Ruling as they thought best under this head.
The people, ignorant of what was done,
Obedience yielded as to Cyrus' son.

Touched with this news, to Persiä he makes;
But in the way his sword just vengeance takes,
Unsheathes, as he his horse mounted on high,
And with a mortal thrust wounds him in the thigh,
Which ends before begun his home-bred war,
So yields to death, that dreadful conqueror.
Grief for his brother's death he did express,
And more because he diéd issueless.
The male line of great Cyrus now had end;
The female to many ages did extend.
A Babylon in Egypt did he make,
And Meroë built for his fair sister's sake.
Eight years he reigned, a short, yet too long, time,
Cut off in wickedness, in strength, and prime.

THE INTERREGNUM BETWEEN CAMBYSES AND DARIUS HYSTASPES.

Childless Cambyses on the sudden dead,
The princes meet to choose one in his stead,
Of which the chief were seven, called satraps,
Who, like to kings, ruled kingdoms as they please;
Descended all of Achæmenes' blood,
And kinsmen in account to the king they stood.
And first these noble Magi agree upon
To thrust the impostor Smerdis out of throne.
Then forces instantly they raise, and rout
The king with his conspirators so stout;

The Four Monarchies

But yet 'fore this was done much blood was shed,
And two of these great peers in field lay dead.
Some write that, sorely hurt, they escaped away;
But so or no, sure 't is they won the day.
All things in peace, and rebels throughly quelled,
A consultation by those states was held
What form of government now to erect,
The old or new, which best, in what respect.
The greater part declined a monarchy,
So late crushed by their prince's tyranny,
And thought the people would more happy be
If governed by an aristocracy.
But others thought — none of the dullest brain —
That better one than many tyrants reign.
What arguments they used I know not well, —
Too politic, it 's like, for me to tell, —
But in conclusiön they all agree
Out of the seven a monarch chosen be.
All envy to avoid, this was thought on:
Upon a green to meet by rising sun,
And he whose horse before the rest should neigh
Of all the peers should have precedency.
They all attend on the appointed hour,
Praying to fortune for a kingly power;
Then mounting on their snorting coursers proud,
Darius' lusty stallion neighed full loud.
The nobles all alight, bow to their king,
And joyful acclamations shrill they ring.

A thousand times "Long live the king!" they cry;
"Let tyranny with dead Cambyses die!"
Then all attend him to his royal room.
Thanks for all this to his crafty stable-groom.

DARIUS HYSTASPES.

Darius by election made a king,
His title to make strong omits no thing:
He two of Cyrus' daughters then doth wed,
Two of his nieces takes to nuptial bed,
By which he cuts their hopes for future time
That by such steps to kingdoms often climb.
And now a king by marriage, choice, and blood,
Three strings to his bow, the least of which is good,
Yet firmly more the people's hearts to bind
Made wholesome, gentle laws which pleased each mind.
His courtesy and affability
Much gained the hearts of his nobility.
Yet, notwithstanding all he did so well,
The Babylonians against their prince rebel.
An host he raised the city to reduce;
But men against those walls were of no use.
Then brave Zopyrus, for his master's good,
His manly face disfigures, spares no blood,
With his own hands cuts off his ears and nose,
And with a faithful fraud to the town he goes,
Tells them how harshly the proud king hath dealt,
That for their sakes his cruelty he felt—

Desiring of the prince to raise the siege,
This violence was done him by his liege.
This told, for entrance there he stood not long,
For they believed his nose more than his tongue.
With all the city's strength they him betrust;
If he command, obey the greatest must.
When opportunity he saw was fit,
Delivers up the town, and all in it.
To lose a nose to win a town 's no shame;
But who dares venture such a stake for the game?
Than thy disgrace thine honor 's manifold,
Who doth deserve a statue made of gold;
Nor can Darius in his monarchy
Scarce find enough to thank thy loyalty.
Yet o'er thy glory we must cast this veil —
Thy craft more than thy valor did prevail.
Darius, in the second of his reign,
An edict for the Jews published again
The temple to rebuild, for that did rest
Since Cyrus' time; Cambyses did molest.
He, like a king, now grants a charter large,
Out of his own revenues bears the charge,
Gives sacrifices, wheat, wine, oil, and salt,
Threats punishment to him that through default
Shall let the work, or keep back anything
Of what is freely granted by the king;
And on all kings he pours out execrations
That shall once dare to raze those firm foundations.

They, thus backed by the king, in spite of foes
Built on and prospered till their house they close,
And in the sixth year of his friendly reign
Set up a temple (though a less) again.
Darius on the Scythians made a war.
Entering that large and barren country far,
A bridge he made, which served for boat and barge
O'er Ister fair, with labor and with charge.
But in that desert, 'mongst his barbarous foes,
Sharp wants, not swords, his valor did oppose:
His army fought with hunger and with cold,
Which to assail his royal camp were bold.
By these alone his host was pinched so sore
He warred defensive, not offensive more.
The savages did laugh at his distress.
Their minds by hieroglyphics they express:
A frog, a mouse, a bird, an arrow, sent.
The king will needs interpret their intent
Possessiön of water, earth, and air;
But wise Gobryas reads not half so fair.
Quoth he, "Like frogs in water we must dive,
Or like to mice under the earth must live,
Or fly like birds in unknown ways full quick,
Or Scythian arrows in our sides must stick."
The king, seeing his men and victuals spent,
This fruitless war began late to repent,
Returned with little honor, and less gain,
His enemies scarce seen, then much less slain.

The Four Monarchies

He after this intends Greece to invade,
But troubles in Less Asiā him stayed,
Which hushed, he straight so orders his affairs
For Attica an army he prepares,
But, as before, so now, with ill success,
Returned with wondrous loss, and honorless.
Athens, perceiving now her desperate state,
Armed all she could, which eleven thousand made;
By brave Miltiades, their chief, being led,
Darius' multitudes before them fled.
At Marathon this bloody field was fought,
Where Grecians proved themselves right soldiers stout.
The Persians to their galleys post with speed,
Where an Athenian showed a valiant deed —
Pursues his flying foes then on the sand,
He stays a launching galley with his hand,
Which soon cut off, enraged, he with his left
Renews his hold, and when of that bereft
His whetted teeth he claps in the firm wood;
Off flies his head, down showers his frolic blood.
Go, Persians, carry home that angry piece
As the best trophy which ye won in Greece.
Darius, light, yet heavy, home returns,
And for revenge his heart still restless burns.
His queen, Atossa, caused all this stir
For Grecian maids, 't is said, to wait on her.
She lost her aim; her husband, he lost more —
His men, his coin, his honor, and his store,

And the ensuing year ended his life,
'T is thought, through grief of this successless strife.
Thirty-six years this noble prince did reign;
Then to his second son did all remain.

XERXES.

Xerxes, Darius' and Atossa's son,
Grandchild to Cyrus, now sits on the throne
(His eldest brother put beside the place,
Because this was first born of Cyrus' race);
His father not so full of lenity
As was his son of pride and cruelty.
He with his crown receives a double war:
The Egyptians to reduce, and Greece to mar.
The first began and finished in such haste
None write by whom nor how 't was overpast.
But for the last he made such preparation
As if to dust he meant to grind that nation;
Yet all his men and instruments of slaughter
Producéd but derisiön and laughter.
Sage Artabańus' counsel had he taken,
And his cousin, young Mardonius, forsaken,
His soldiers, credit, wealth, at home had stayed,
And Greece such wondrous triumphs ne'er had made.
The first dehorts and lays before his eyes
His father's ill success in his enterprise
Against the Scythians, and Grecians, too;
What infamy to his honor did accrue.

The Four Monarchies

Flattering Mardonius, on the other side,
With conquest of all Europe feeds his pride.
Vain Xerxes thinks his counsel hath most wit
That his ambitious humor best can fit;
And by this choice unwarily posts on
To present loss, future subversiön.
Although he hasted, yet four years were spent
In great provisions for this great intent.
His army of all nations was compounded
That the vast Persian government surrounded.
His foot was seventeen hundred thousand strong;
Eight hundred thousand horse to these belong.
His camels, beasts for carriage, numberless,
For truth's ashamed how many to express.
The charge of all he severally commended
To princes of the Persian blood descended;
But the command of these commanders all
Unto Mardonius, made their general.
He was the son of the forenamed Gobryas,
Who marriéd the sister of Darius.
Such his land forces were. Then next a fleet
Of two and twenty thousand galleys meet,
Manned with Phenicians and Pamphylians,
Cypriotes, Dorians, and Cilicians,
Lycians, Carians, and Ionians,
Æolians, and the Hellespontines;
Besides the vessels for his transportation,
Which to three thousand came, by best relation.

Brave Artemisia, Halicarnassus' queen,
In person present for his aid was seen,
Whose galleys all the rest in neatness pass
Save the Sidonians, where Xerxes was.
But hers she kept still separate from the rest,
For to command alone she judged was best.
O noble queen, thy valor I commend;
But pity 't was thine aid thou here didst lend.
At Sardis, in Lydia, all these do meet,
Whither rich Pythius comes Xerxes to greet,
Feasts all this multitude of his own charge,
Then gives the king a king-like gift full large —
Three thousand talents of the purest gold,
Which mighty sum all wondered to behold.
Then humbly to the king he makes request
One of his five sons there might be released
To be to his age a comfort and a stay;
The other four he freely gave away.
The king calls for the youth, who being brought,
Cuts him in twain for whom his sire besought;
Then laid his parts on both sides of the way,
'Twixt which his soldiers marched in good array.
For his great love is this thy recompense?
Is this to do like Xerxes or a prince?
Thou shame of kings, of men the detestation,
I rhetoric want to pour out execration.
First thing he did that's worthy of recount
A sea-passage cut behind Athos' mount.

The Four Monarchies

Next o'er the Hellespont a bridge he made
Of boats together coupled and there laid.
But winds and waves those iron bands did break;
To cross the sea such strength he found too weak;
Then whips the sea, and with a mind most vain
He fetters casts therein the same to chain;
The workmen put to death the bridge that made
Because they wanted skill the same to have stayed.
Seven thousand galleys chained by Tyrians' skill
Firmly at last accomplishéd his will.
Seven days and nights his host, without least stay,
Was marching o'er this new deviséd way.
Then in Abydos' plains mustering his forces,
He glories in his squadrons and his horses;
Long viewing them, thought it great happiness
One king so many subjects should possess.
But yet this sight from him producéd tears
That none of those could live an hundred years.
What after did ensue, had he foreseen,
Of so long time his thoughts had never been.
Of Artabanus he again demands
How of this enterprise his thought now stands.
His answer was, both sea and land he feared;
Which was not vain, as after soon appeared.
But Xerxes resolute to Thrace goes first.
His host all Lissus drinks to quench its thirst;
And for his cattle all Pissirus' lake
Was scarce enough for each a draught to take.

Then, marching on to the strait Thermopylæ,
The Spartan meets him, brave Leonide.
This 'twixt the mountains lies, half acre wide,
That pleasant Thessaly from Greece divide.
Two days and nights a fight they there maintain,
Till twenty thousand Persians fell down slain;
And all that army then, dismayed, had fled
But that a fugitive discoveréd
How some might o'er the mountains go about
And wound the backs of those brave warriors stout.
They thus, behemmed with multitude of foes,
Laid on more fiercely their deep mortal blows.
None cries for quarter, nor yet seeks to run,
But on their ground they die, each mother's son.
O noble Greeks, how now degenerate!
Where is the valor of your ancient state
Whenas one thousand could a million daunt?
Alas, it is Leonidas you want!
This shameful victory cost Xerxes dear;
Among the rest, two brothers he lost there.
And as at land, so he at sea was crossed:
Four hundred stately ships by storms were lost;
Of vessels small almost innumerable,
The harbors to contain them were not able.
Yet, thinking to outmatch his foes at sea,
Inclosed their fleet in the strait of Eubœa;
But they, as fortunate at sea as land,
In this strait, as the other, firmly stand,

And Xerxes' mighty galleys battered so
That their split sides witnessed his overthrow.
Then in the strait of Salamis he tried
If that small number his great force could bide;
But he, in daring of his forward foe,
Receivéd there a shameful overthrow.
Twice beaten thus at sea, he warred no more,
But then the Phocians' country wasted sore.
They no way able to withstand his force,
The brave Themistocles takes this wise course:
In secret manner word to Xerxes sends
That Greeks to break his bridge shortly intend;
And, as a friend, warns him, whate'er he do,
For his retreat to have an eye thereto.
He, hearing this, his thoughts and course home bended,
Much fearing that which never was intended.
Yet 'fore he went, to help out his expense,
Part of his host to Delphos sent from thence
To rob the wealthy temple of Apollo.
But mischief sacrilege doth ever follow.
Two mighty rocks brake from Parnassus' hill,
And many thousands of those men did kill;
Which accident the rest affrighted so
With empty hands they to their master go.
He, finding all to tend to his decay,
Fearing his bridge, no longer there would stay.
Three hundred thousand yet he left behind
With his Mardonius, index of his mind;

Who, for his sake, he knew would venture far,
Chief instigator of this hapless war.
He instantly to Athens sends for peace,
That all hostility from thenceforth cease,
And that with Xerxes they would be at one;
So should all favor to their state be shown.
The Spartans, fearing Athens would agree,
As had Macedon, Thebes, and Thessaly,
And leave them out this shock now to sustain,
By their ambassador they thus complain
That Xerxes' quarrel was 'gainst Athens' state,
And they had helped them as confederate;
If in their need they should forsake their friend,
Their infamy would last till all things end.
But the Athenians this peace detest,
And thus replied unto Mardon's request:
That while the sun did run his endless course
Against the Persians they would bend their force;
Nor could the brave ambassador he sent
With rhetoric gain better compliment —
A Macedonian born, and great commander,
No less than grandsire to great Alexander.
Mardonius proud, hearing this answer stout,
To add more to his numbers lays about;
And of those Greeks which by his skill he won
He fifty thousand joins unto his own.
The other Greeks which were confederate
In all one hundred and ten thousand made.

The Athenians could but forty thousand arm,
The rest had weapons would do little harm;
But that which helped defects and made them bold
Was victory by oracle foretold.
Then for one battle shortly all provide
Where both their controversies they 'll decide.
Ten days these armies did each other face.
Mardonius, finding victuals waste apace,
No longer dared, but bravely onset gave.
The other not a hand or sword would wave
Till in the entrails of their sacrifice
The signal of their victory did rise;
Which found, like Greeks they fight, the Persians fly,
And troublesome Mardonius now must die.
All 's lost; and of three hundred thousand men
Three thousand only can run home again.
For pity let those few to Xerxes go
To certify his final overthrow.
Same day the small remainder of his fleet
The Grecians at Mycale in Asia meet,
And there so utterly they wrecked the same
Scarce one was left to carry home the fame.
Thus did the Greeks consume, destroy, disperse,
That army which did fright the universe.
Scorned Xerxes, hated for his cruelty,
Yet ceases not to act his villainy.
His brother's wife solicits to his will;
The chaste and beauteous dame refused still.

Some years by him in this vain suit were spent,
Nor prayers nor gifts could win him least content,
Nor matching of her daughter to his son;
But she was still as when he first begun.
When jealous Queen Amestris of this knew
She harpy-like upon the lady flew,
Cut off her breasts, her lips, her nose, and ears,
And leaves her thus besmeared in blood and tears.
Straight comes her lord, and finds his wife thus lie.
The sorrow of his heart did close his eye.
He dying to behold that wounding sight
Where he had sometime gazed with great delight,
To see that face where rose and lilies stood
O'erflown with torrents of her guiltless blood,
To see those breasts where chastity did dwell
Thus cut and mangled by a hag of hell,
With laden heart unto the king he goes,
Tells as he could his unexpresséd woes.
But for his deep complaints and showers of tears
His brother's recompense was naught but jeers.
The grievéd prince, finding nor right nor love,
To Bactria his household did remove.
His brother sent soon after him a crew
Which him and his most barbarously there slew.
Unto such height did grow his cruelty,
Of life no man had least security.
At last his uncle did his death conspire,
And for that end his eunuch he did hire,

Who privately him smothered in his bed,
But yet by search he was found murderéd.
Then Artabanus, hirer of this deed,
That from suspiciön he might be freed,
Accused Darius, Xerxes' eldest son,
To be the author of the crime was done,
And by his craft ordered the matter so
That the prince, innocent, to death did go.
But in short time this wickedness was known,
For which he diëd, and not he alone,
But all his family was likewise slain.
Such justice in the Persian court did reign.
The eldest son thus immaturely dead,
The second was enthroned in his father's stead.

ARTAXERXES LONGIMANUS.

Amongst the monarchs next this prince had place,
The best that ever sprung of Cyrus' race.
He first war with revolted Egypt made,
To whom the perjured Grecians lent their aid
Although to Xerxes they not long before
A league of amity had firmly swore,
Which had they kept, Greece had more nobly done
Than when the world they after overrun.
Greeks and Egyptians both he overthrows,
And pays them both according as he owes.
Which done, a sumptuous feast makes like a king,
Where ninescore days are spent in banqueting;

His princes, nobles, and his captains calls
To be partakers of these festivals.
His hangings white and green, and purple dye,
With gold and silver beds most gorgeously.
The royal wine in golden cups did pass;
To drink more than he list none bidden was.
Queen Vashti also feasts; but 'fore 't is ended
She 's from her royalty, alas, suspended,
And one more worthy placéd in her room;
By Memucan's advice so was the doom.
What Esther was and did, the story read,
And how her countrymen from spoil she freed;
Of Haman's fall, and Mordecai's great rise,
The might of the prince, the tribute of the isles.
Good Ezra in the seventh year of his reign
Did for the Jews commission large obtain,
With gold and silver, and whate'er they need;
His bounty did Darius' far exceed.
And Nehemiah, in his twentieth year,
Went to Jerusalem, his city dear,
Rebuilt those walls which long in rubbish lay,
And o'er his opposites still got the day.
Unto this king Themistocles did fly
When under ostracism he did lie —
For such ingratitude did Athens show
This valiant knight, whom they so much did owe.
Such royal bounty from his prince he found
That in his loyalty his heart was bound.

The king not little joyful of this chance,
Thinking his Grecian wars now to advance,
And for that end great preparation made
Fair Attica a third time to invade.
His grandsire's old disgrace did vex him sore,
His father Xerxes' loss and shame much more.
For punishment their breach of oath did call
This noble Greek, now fit for general.
Provisions then and season being fit,
To Themistocles this war he doth commit,
Who for his wrong he could not choose but deem
His country nor his friends would much esteem;
But he all injury had soon forgot,
And to his native land could bear no hate,
Nor yet disloyal to his prince would prove,
By whom obliged by bounty and by love.
Either to wrong did wound his heart so sore
To wrong himself by death he chose before.
In this sad conflict marching on his ways,
Strong poison took, so put an end to his days.
The king, this noble captain having lost,
Dispersed again his newly-levied host.
Rest of his time in peace he did remain,
And died the two and fortieth of his reign.

DARIUS NOTHUS.

Three sons great Artaxerxes left behind;
The eldest to succeed, that was his mind.

His second brother with him fell at strife,
Still making war till first had lost his life.
Then the survivor is by Nothus slain,
Who now sole monarch doth of all remain.
The first two sons are by historians thought
By fair Queen Esther to her husband brought.
If so they were, the greater was her moan
That for such graceless wretches she did groan.
Revolting Egypt 'gainst this king rebels,
His garrison drives out that mongst them dwells;
Joins with the Greeks, and so maintains their right
For sixty years, maugre the Persians' might.
A second trouble after this succeeds,
Which from remissness in Less Asia breeds.
Amorges, whom for viceroy he ordained,
Revolts, treasure and people having gained,
Plunders the country, and much mischief wrought
Before things could to quietness be brought.
The king was glad with Sparta to make peace,
That so he might those troubles soon appease;
But they in Asiä must first restore
All towns held by his ancestors before.
The king much profit reapéd by this league,
Regains his own, then doth the rebel break,
Whose strength by Grecians' help was overthrown,
And so each man again possessed his own.
This king, Cambyses-like, his sister wed,
To which his pride more than his lust him led;

For Persian kings then deemed themselves so good
No match was high enough but their own blood.
Two sons she bore, the youngest Cyrus named,
A prince whose worth by Xenophon is famed.
His father would no notice of that take,
Prefers his brother for his birthright's sake.
But Cyrus scorns his brother's feeble wit,
And takes more on him than was judgéd fit.
The king, provoked, sends for him to the court,
Meaning to chastise him in sharpest sort;
But in his slow approach ere he came there
His father died, so put an end to his fear.
About nineteen years this Nothus reigned, which run,
His large dominions left to his eldest son.

ARTAXERXES MNEMON.

Mnemon now sat upon his father's throne,
Yet fears all he enjoys is not his own;
Still on his brother casts a jealous eye,
Judging his actions tend to his injury.
Cyrus, on the other side, weighs in his mind
What help in his enterprise he's like to find.
His interest in the kingdom, now next heir,
More dear to his mother than his brother far,
His brother's little love like to be gone,
Held by his mother's intercessiön —
These and like motives hurry him amain
To win by force what right could not obtain;

And thought it best now in his mother's time
By lower steps toward the top to climb.
If in his enterprise he should fall short,
She to the king would make a fair report;
He hoped if fraud nor force the crown would gain
Her prevalence a pardon might obtain.
From the lieutenant first he takes away
Some towns, commodious in Less Asia,
Pretending still the profit of the king,
Whose rents and customs duly he sent in.
The king, finding revenues now amended,
For what was done seeméd no whit offended.
Then next he takes the Spartans into pay—
One Greek could make ten Persians run away.
Great care was his pretense those soldiers stout
The rovers in Pisidia should drive out;
But lest some blacker news should fly to court
Prepares himself to carry the report,
And for that end five hundred horse he chose.
With posting speed on toward the king he goes.
But fame, more quick, arrives ere he comes there,
And fills the court with tumult and with fear.
The old queen and the young at bitter jars,
The last accused the first for these sad wars;
The wife against the mother still doth cry
To be the author of conspiracy.
The king, dismayed, a mighty host doth raise,
Which Cyrus hears, and so foreslows his pace;

The Four Monarchies

But as he goes his forces still augments,—
Seven hundred Greeks repair for his intents,
And others to be warmed by this new sun
In numbers from his brother daily run.
The fearful king at last musters his forces,
And counts nine hundred thousand foot and horses.
Three hundred thousand he to Syria sent
To keep those straits his brother to prevent.
Their captain, hearing but of Cyrus' name,
Forsook his charge, to his eternal shame.
This place so made by nature and by art
Few might have kept it had they had a heart.
Cyrus despaired a passage there to gain,
So hired a fleet to waft him o'er the main.
The amazéd king was then about to fly
To Bactria, and for a time there lie,
Had not his captains, sore against his will,
By reason and by force detained him still.
Up then with speed a mighty trench he throws
For his security against his foes,
Six yards the depth and forty miles in length,
Some fifty or else sixty foot in breadth;
Yet for his brother's coming durst not stay,—
He safest was when farthest out of the way.
Cyrus, finding his camp and no man there,
Was not a little jocund at his fear.
On this he and his soldiers careless grow,
And here and there in carts their arms they throw,

When suddenly their scouts come in and cry,
"Arm! Arm! The king with all his host is nigh!"
In this confusion, each man as he might
Gets on his arms, arrays himself for fight,
And rangéd stood by great Euphrates' side
The brunt of that huge multitude to abide,
Of whose great numbers their intelligence
Was gathered by the dust that rose from thence,
Which like a mighty cloud darkened the sky,
And black and blacker grew as they drew nigh.
But when their order and their silence saw,
That more than multitudes their hearts did awe;
For tumult and confusion they expected,
And all good discipline to be neglected.
But long under their fears they did not stay,
For at first charge the Persians ran away,
Which did such courage to the Grecians bring
They all adoréd Cyrus for their king;
So had he been, and got the victory,
Had not his too much valor put him by.
He with six hundred on a squadron set
Of thousands six wherein the king was yet,
And brought his soldiers on so gallantly
They ready were to leave their king and fly;
Whom Cyrus spies, cries loud, "I see the man!"
And with a full career at him he ran.
And in his speed a dart him hit in the eye;—
Down Cyrus falls, and yields to destiny.

His host in chase know not of this disaster,
But tread down all so to advance their master;
But when his head they spy upon a lance,
Who knows the sudden change made by this chance?
Senseless and mute they stand, yet breathe out groans,
Nor Gorgon's head like this transformed to stones.
After this trance revenge new spirits blew,
And now more eagerly their foes pursue,
And heaps on heaps such multitudes they laid
Their arms grew weary by their slaughters made.
The king unto a country village flies,
And for a while unkingly there he lies;
At last displays his ensign on a hill,
Hoping by that to make the Greeks stand still,
But was deceived. To him they run amain;
The king upon the spur runs back again.
But they, too faint still to pursue their game,
Being victors oft now to their camp they came,
Nor lacked they any of their number small,
Nor wound received but one among them all.
The king, with his dispersed, also encamped,
With infamy upon each forehead stamped.
His hurried thoughts he after re-collects;
Of this day's cowardice he fears the effects.
If Greeks in their own country should declare
What dastards in the field the Persians are,
They in short time might place one on his throne,
And rob him both of scepter and of crown.

To hinder their return by craft or force
He judged his wisest and his safest course;
Then sends that to his tent they straight address,
And there all wait his mercy weaponless.
The Greeks with scorn reject his proud commands,
Asking no favor where they feared no bands.
The troubled king his herald sends again,
And sues for peace, that they his friends remain.
The smiling Greeks reply they first must bait;
They were too hungry to capitulate.
The king great store of all provision sends,
And courtesy to the utmost he pretends;
Such terror on the Persians then did fall
They quaked to hear them to each other call.
The king, perplexed, there dares not let them stay,
And fears as much to let them march away.
But kings ne'er want such as can serve their will,
Fit instruments to accomplish what is ill;
As Tissaphernes, knowing his master's mind,
Their chief commanders feasts, and yet more kind,
With all the oaths and deepest flattery,
Gets them to treat with him in privacy,
But violates his honor and his word,
And villain-like there puts them all to the sword.
The Greeks, seeing their valiant captains slain,
Chose Xenophon to lead them home again.
But Tissaphernes what he could devise
Did stop the way in this their enterprise;

The Four Monarchies

But when through difficulties all they brake,
The country burned they no relief might take.
But on they march, through hunger and through cold,
O'er mountains, rocks, and hills, as lions bold;
Nor rivers' course nor Persians' force could stay,
But on to Trebizond they kept their way.
There was of Greeks settled a colony,
Who after all received them joyfully.
Thus finishing their travail, danger, pain,
In peace they saw their native soil again.
The Greeks now, as the Persian king suspected,
The Asiatics' cowardice detected —
The many victories themselves did gain,
The many thousand Persians they had slain,
And how their nation with facility
Might gain the universal monarchy.
They then Dercyllidas send with an host,
Who with the Spartans on the Asian coast
Town after town with small resistance takes,
Which rumor makes great Artaxerxes quake.
The Greeks by this success encouraged so,
Their king Agesilaus doth over go.
By Tissaphernes he 's encounteréd,
Lieutenant to the king; but soon he fled,
Which overthrow incensed the king so sore
That Tissaphern must be viceroy no more.
Tithraustes then is placéd in his stead,
Commission hath to take the other's head;

Of that perjurious wretch this was the fate,
Whom the old queen did bear a mortal hate.
Tithraustes trusts more to his wit than arms,
And hopes by craft to quit his master's harms.
He knows that many a town in Greece envies
The Spartan state, which now so fast did rise;
To them he thirty thousand talents sent,
With suit their arms against their foes be bent.
They to their discontent receiving hire,
With broils and quarrels set all Greece on fire.
Agesilaus is called home with speed.
To defend, more than offend, there was need.
Their winnings lost, and peace they're glad to take
On such conditions as the king will make.
Dissension in Greece continued so long
Till many a captain fell, both wise and strong,
Whose courage naught but death could ever tame.
'Mongst these Epaminondas wants no fame,
Who had, as noble Raleigh doth evince,
All the peculiar virtues of a prince.
But let us leave these Greeks to discord bent,
And turn to Persia, as is pertinent.
The king, from foreign parts now well at ease,
His home-bred troubles sought how to appease.
The two queens by his means seem to abate
Their former envy and inveterate hate;
But the old queen, implacable in strife,
By poison caused the young to lose her life.

The king, highly enraged, doth thereupon
From court exile her unto Babylon;
But shortly calls her home, her counsels prize,
A lady very wicked, but yet wise.
Then in voluptuousness he leads his life,
And weds his daughter for a second wife.
But long in ease and pleasure did not lie;
His sons sore vexed him by disloyalty.
Such as would know at large his wars and reign,
What troubles in his house he did sustain,
His match incestuous, cruelties of the queen,
His life may read in Plutarch to be seen.
Forty-three years he ruled, then turned to dust,
A king nor good, nor valiant, wise, nor just.

DARIUS OCHUS.

Ochus, a wicked and rebellious son,
Succeeds in the throne, his father being gone.
Two of his brothers in his father's days,
To his great grief, most subtilely he slays;
And, being king, commands those that remain
Of brethren and of kindred to be slain.
Then raises forces, conquers Egypt land,
Which in rebellion sixty years did stand,
And in the twenty-third of his cruel reign
Was by his eunuch, the proud Bagoas, slain.

ARSAMES, OR ARSES.

Arsames, placed now in his father's stead
By him that late his father murderéd,
Some write that Arsames was Ochus' brother,
Enthroned by Bagoas in the room of the other;
But why his brother 'fore his son succeeds
I can no reason give, 'cause none I read.
His brother, as 't is said, long since was slain,
And scarce a nephew left that now might reign.
What acts he did time hath not now left penned,
But most suppose in him did Cyrus end,
Whose race long time had worn the diadem,
But now 's devolvéd to another stem.
Three years he reigned, then drank his father's cup
By the same eunuch who first set him up.

DARIUS CODOMANNUS.

Darius, by this Bagoas set in throne,—
Complotter with him in the murder done,—
He was no sooner settled in his reign
But Bagoas falls to his practices again,
And the same sauce had servéd him, no doubt,
But that his treason timely was found out;
And so this wretch, a punishment too small,
Lost but his life for his horrid treasons all.
This Codomannus now upon the stage
Was to his predecessors chamber-page.

Some write great Cyrus' line was not yet run,
But from some daughter this new king was sprung.
If so or not we cannot tell, but find
That several men will have their several mind.
Yet in such differences we may be bold
With the learnéd and judicious still to hold;
And this 'mongst all 's no controverted thing,
That this Darius was last Persian king,
Whose wars and losses we may better tell
In Alexander's reign, who did him quell:
How from the top of world's felicity
He fell to depths of greatest misery;
Whose honors, treasures, pleasures, had short stay —
One deluge came and swept them all away,
And in the sixth year of his hapless reign
Of all did scarce his winding-sheet remain;
And last, a sad catastrophe to end,
Him to the grave did traitor Bessus send.

The End of the Persian Monarchy.

THE THIRD MONARCHY, BEING THE GRECIAN, BEGINNING UNDER ALEXANDER THE GREAT IN THE ONE HUNDRED AND TWELFTH OLYMPIAD.

Great Alexander was wise Philip's son,
He to Amyntas, kings of Macedon;
The cruel, proud Olympias was his mother,
She to Epirus' warlike king was daughter.
This prince, his father by Pausanias slain,
The twenty-first of his age began to reign.
Great were the gifts of nature which he had,
His education much to those did add;
By art and nature both he was made fit
To accomplish that which long before was writ.
The very day of his nativity
To the ground was burned Diana's temple high,
An omen of their near approaching woe
Whose glory to the earth this king did throw.
His rule to Greece he scorned should be confined;
The universe scarce bound his proud, vast mind.
This is the he-goat which from Grecia came,
That ran in choler on the Persian ram,
That brake his horns, that threw him on the ground;
To save him from his might no man was found.
Philip on this great conquest had an eye,
But death did terminate those thoughts so high;
The Greeks had chose him captain-general,
Which honor to his son did now befall.

The Four Monarchies

For as world's monarch now we speak not on,
But as the king of little Macedon.
Restless both day and night his heart then was
His high resolves which way to bring to pass;
Yet for a while in Greece he's forced to stay,
Which makes each moment seem more than a day.
Thebes and stiff Athens both 'gainst him rebel;
Their mutinies by valor doth he quell.
This done, against both right and nature's laws
His kinsmen put to death, who gave no cause,
That no rebellion in his absence be,
Nor making title unto sovereignty;
And all whom he suspects or fears will climb
Now taste of death, lest they deserve it in time.
Nor wonder is it if he in blood begin,
For cruelty was his parental sin.
Thus easéd now of troubles and of fears,
Next spring his course to Asiä he steers;
Leaves sage Antipater at home to sway,
And through the Hellespont his ships made way.
Coming to land, his dart on shore he throws,
Then with alacrity he after goes;
And with a bounteous heart and courage brave
His little wealth among his soldiers gave.
And being asked what for himself was left,
Replied, Enough, sith only hope he kept.
Thirty-two thousand made up his foot force,
To which were joined five thousand goodly horse.

Then on he marched. In his way he viewed old Troy,
And on Achilles' tomb with wondrous joy
He offered, and for good success did pray
To him, his mother's ancestor, men say.
When news of Alexander came to court,
To scorn at him Darius had good sport;
Sends him a frothy and contemptuous letter:
Styles him disloyal servant, and no better;
Reproves him for his proud audacity
To lift his hand 'gainst such a monarchy.
Then to his lieutenant he in Asia sends
That he be taken alive, for he intends
To whip him well with rods, and so to bring
That boy so malapert before the king.
Ah, fond, vain man, whose pen ere while
In lower terms was taught a higher style!
To river Granicus Alexander hies,
Which in Phrygia near Propontis lies.
The Persians ready for encounter stand,
And strive to keep his men from off the land;
Those banks so steep the Greeks yet scramble up,
And beat the coward Persians from the top,
And twenty thousand of their lives bereave,
Who in their backs did all their wounds receive.
This victory did Alexander gain
With loss of thirty-four of his there slain.
Then Sardis he, and Ephesus, did gain,
Where stood of late Diana's wondrous fane.

And by Parmenio, of renownéd fame,
Miletus and Pamphylia overcame;
Halicarnassus and Pisidia
He for his master takes, with Lycia.
Next Alexander marched toward the Black Sea,
And easily takes old Gordium in his way,
Of ass-eared Midas once the regal seat,
Whose touch turned all to gold, yea, e'en his meat.
There the prophetic knot he cuts in twain,
Which whoso doth must lord of all remain.
Now news of Memnon's death, the king's viceroy,
To Alexander's heart's no little joy,
For in that peer more valor did abide
Than in Darius' multitude beside.
In his stead was Arses placed, but durst not stay,
Yet set one in his room, and ran away;
His substitute, as fearful as his master,
Runs after, too, and leaves all to disaster.
Then Alexander all Cilicia takes,
No stroke for it he struck, their hearts so quake.
To Greece he thirty thousand talents sends
To raise more force to further his intents.
Then o'er he goes Darius now to meet,
Who came with thousand thousands at his feet—
Though some there be, perhaps more likely, write
He but four hundred thousand had to fight;
The rest attendants, which made up no less,
Both sexes there were almost numberless.

For this wise king had brought, to see the sport,
With him the greatest ladies of the court,
His mother, his beauteous queen and daughters,
It seems, to see the Macedonian slaughters.
It's much beyond my time and little art
To show how great Darius played his part,
The splendor and the pomp he marchéd in,
For since the world was no such pageant seen.
Sure 't was a goodly sight there to behold
The Persians clad in silk and glistering gold,
The stately horses trapped, the lances gilt,
As if addressed now all to run a tilt.
The holy fire was borne before the host,
For sun and fire the Persians worship most;
The priests, in their strange habit, follow after,
An object not so much of fear as laughter.
The king sat in a chariot made of gold,
With crown and robes most glorious to behold,
And o'er his head his golden gods on high
Support a party-colored canopy.
A number of spare horses next were led,
Lest he should need them in his chariot's stead;
But those that saw him in this state to lie
Supposed he neither meant to fight nor fly.
He fifteen hundred had like women dressed,
For thus to fright the Greeks he judged was best;
Their golden ornaments how to set forth
Would ask more time than were their bodies worth.

The Four Monarchies

Great Sisygambis she brought up the rear;
Then such a world of wagons did appear,
Like several houses moving upon wheels,
As if she'd drawn whole Shushan at her heels.
This brave virago to the king was mother,
And as much good she did as any other.
Now lest this gold and all this goodly stuff
Had not been spoil and booty rich enough,
A thousand mules and camels ready wait
Laden with gold, with jewels, and with plate.
For sure Darius thought at the first sight
The Greeks would all adore but none would fight.
But when both armies met, he might behold
That valor was more worth than pearls or gold,
And that his wealth served but for baits to allure
To make his overthrow more fierce and sure.
The Greeks came on, and with a gallant grace
Let fly their arrows in the Persians' face.
The cowards, feeling this sharp, stinging charge,
Most basely ran, and left their king at large,
Who from his golden coach is glad to alight
And cast away his crown for swifter flight.
Of late like some immovable he lay;
Now finds both legs and horse to run away.
Two hundred thousand men that day were slain,
And forty thousand prisoners also ta'en,
Besides the queens and ladies of the court,
If Curtiüs be true in his report.

The regal ornaments were lost, the treasure
Divided at the Macedonian's pleasure.
Yet all this grief, this loss, this overthrow,
Was but beginning of his future woe.
The royal captives brought to Alexander,
Toward them demeaned himself like a commander;
For though their beauties were unparalleled,
Conquered himself now he who 'd conqueréd,
Preserved their honor, used them bounteously,
Commands no man should do them injury;—
And this to Alexander is more fame
Than that the Persian king he overcame.
Two hundred eighty Greeks he lost in fight
By too much heat, not wounds, as authors write.
No sooner had this victor won the fields
But all Phenicia to his pleasure yields,
Of which the government he doth commit
Unto Parmenio, of all most fit.
Darius, now less lofty than before,
To Alexander writes he would restore
Those mournful ladies from captivity,
For whom he offers him a ransom high
But down his haughty stomach could not bring
To give this conqueror the style of king.
This letter Alexander doth disdain,
And in short terms sends this reply again:
A king he was, and that not only so,
But of Darius king, as he should know.

The Four Monarchies

Next Alexander unto Tyre doth go.
His valor and his victories they know;
To gain his love the Tyrians intend,
Therefore a crown and great provision send.
Their present he receives with thankfulness,
Desires to offer unto Hercules,
Protector of their town, by whom defended,
And from whom he lineally descended.
But they accept not this in any wise
Lest he intend more fraud than sacrifice;
Sent word that Hercules his temple stood
In the old town, which then lay like a wood.
With this reply he was so deep enraged
To win the town his honor he engaged;
And now, as Babel's king did once before,
He leaves not till he made the sea firm shore.
But far less time and cost he did expend —
The former ruins forwarded his end;
Moreover, he 'd a navy at command,
The other by his men fetched all by land.
In seven months' time he took that wealthy town,
Whose glory now a second time 's brought down.
Two thousand of the chief he crucified,
Eight thousand by the sword then also died,
And thirteen thousand galley-slaves he made;
And thus the Tyrians for mistrust were paid.
The rule of this he to Philotas gave,
Who was the son of that Parmenio brave.

Ciliciā to Socrates doth give,
For now 's the time captains like kings may live.
Sidon he on Hephæstiön bestows,
For that which freely comes as freely goes;
He scorns to have one worse than had the other,
So gives his little lordship to another.
Hephæstion, having chief command of the fleet,
At Gaza now must Alexander meet.
Darius, finding troubles still increase,
By his ambassadors now sues for peace,
And lays before great Alexander's eyes
The dangers, difficulties, like to rise:
First at Euphrates what he is like to abide,
And then at Tigris' and Araxes' side;
These he may escape, and if he so desire
A league of friendship make firm and entire.
His eldest daughter he in marriage proffers,
And a most princely dowry with her offers —
All those rich kingdoms large that do abide
Betwixt the Hellespont and Halys' side.
But he with scorn his courtesy rejects,
And the distresséd king no whit respects;
Tells him these proffers great in truth were none,
For all he offers now was but his own.
But quoth Parmenio, that brave commander,
"Were I as great as is great Alexander
Darius' offers I would not reject,
But the kingdoms and the lady soon accept."

The Four Monarchies

To which proud Alexander made reply,
"And so, if I Parmenio were, would I."
He now to Gaza goes, and there doth meet
His favorite Hephæstion with his fleet,
Where valiant Betis stoutly keeps the town,
A loyal subject to Darius' crown.
For more repulse the Grecians here abide
Than in the Persian monarchy beside;
And by these walls so many men were slain
That Greece was forced to yield supply again.
But yet this well defended town was taken,
For 't was decreed that empire should be shaken.
Thus Betis taken had holes bored through his feet,
And by command was drawn through every street
To imitate Achilles in his shame,
Who did the like to Hector, of more fame.
What! hast thou lost thy magnanimity?
Can Alexander deal thus cruelly?
Sith valor with heroics is renowned
Though in an enemy it should be found,
If of thy future fame thou hadst regard
Why didst not heap up honors and reward?
From Gaza to Jerusalem he goes,
But in no hostile way, as I suppose.
Him in his priestly robes high Jaddua meets,
Whom with great reverence Alexander greets;
The priest shows him good Daniel's prophecy,
How he should overthrow this monarchy,

By which he was so much encouragéd
No future dangers he did ever dread.
From thence to fruitful Egypt marched with speed,
Where happily in his wars he did succeed;
To see how fast he gained was no small wonder,
For in few days he brought that kingdom under.
Then to the fane of Jupiter he went,
To be installed a god was his intent;
The pagan priest, through hire, or else mistake,
The son of Jupiter did straight him make.
He diabolical must needs remain
That his humanity will not retain.
Thence back to Egypt goes, and in few days
Fair Alexandria from the ground doth raise.
Then settling all things in Less Asiä,
In Syria, Egypt, and Pheniciä,
Unto Euphrates marched and over goes,
For no man's there his army to oppose.
Had Betis but been there now with his band,
Great Alexander had been kept from land.
But as the king, so is the multitude,
And now of valor both are destitute.
Yet he, poor prince, another host doth muster
Of Persians, Scythians, Indians, in a cluster,
Men but in shape and name, of valor none,
Most fit to blunt the swords of Macedon.
Two hundred fifty thousand, by account,
Of horse and foot his army did amount.

The Four Monarchies

For in his multitudes his trust still lay,
But on their fortitude he had small stay;
Yet had some hope that on the spacious plain
His numbers might the victory obtain.
About this time Darius' beauteous queen,
Who had sore travail and much sorrow seen,
Now bids the world adieu, with pain being spent,
Whose death her lord full sadly did lament.
Great Alexander mourns as well as he,
The more because not set at liberty.
When this sad news at first Darius hears,
Some injury was offeréd, he fears;
But when informed how royally the king
Had uséd her and hers in everything,
He prays the immortal gods they would reward
Great Alexander for this good regard;
And if they down his monarchy will throw,
Let them on him this dignity bestow.
And now for peace he sues, as once before,
And offers all he did and kingdoms more,
His eldest daughter for his princely bride,—
Nor was such match in all the world beside,—
And all those countries which betwixt did lie
Phenician sea and great Euphrates high,
With fertile Egypt, and rich Syria,
And all those kingdoms in Less Asiä,
With thirty thousand talents to be paid
For the queen mother and the royal maid;

And till all this be well performed and sure
Ochus his son for hostage should endure.
To this stout Alexander gives no ear,
No, though Parmenio plead, yet will not hear;
Which had he done, perhaps his fame he'd kept,
Nor infamy had waked when he had slept.
For his unlimited prosperity
Him boundless made in vice and cruelty.
Thus to Darius he writes back again:
"The firmament two suns cannot contain;
Two monarchies on earth cannot abide,
Nor yet two monarchs in one world reside."
The afflicted king, finding him set to jar,
Prepares against to-morrow for the war.
Parmenio Alexander wished that night
To force his camp, so vanquish them by flight;
For tumult in the night doth cause most dread,
And weakness of a foe is coveréd.
But he disdained to steal a victory:
The sun should witness of his valor be;
And careless in his bed next morn he lies,
By captains twice he's called before he'll rise.
The armies joined, a while the Persians fight,
And spilled the Greeks some blood before their flight;
But long they stood not ere they're forced to run,
So made an end as soon as well begun.
Forty-five thousand Alexander had,
But 't is not known what slaughter here was made.

The Four Monarchies

Some write the other had a million, some more,
But Quintus Curtius as was said before.
At Arbela this victory was gained,
Together with the town also obtained.
Darius, stripped of all, to Media came,
Accompanied with sorrow, fear, and shame;
At Arbela left ornaments and treasure
Which Alexander deals as suits his pleasure.
This conqueror to Babylon then goes,
Is entertained with joy and pompous shows;
With showers of flowers the streets along are strown,
And incense burned the silver altars on.
The glory of the castle he admires,
The strong foundation and the lofty spires;
In this a world of gold and treasure lay
Which in few hours was carried all away.
With greedy eyes he views this city round
Whose fame throughout the world was so renowned,
And to possess he counts no little bliss
The towers and bowers of proud Semiramis;
Though worn by time, and razed by foes full sore,
Yet old foundations showed, and somewhat more.
With all the pleasures that on earth are found
This city did abundantly abound,
Where four and thirty days he now did stay
And gave himself to banqueting and play.
He and his soldiers wax effeminate,
And former discipline begin to hate.

Whilst reveling at Babylon he lies,
Antipater from Greece sends fresh supplies.
He then to Shushan goes with his new bands,
But needs no force; 't is rendered to his hands.
He likewise here a world of treasure found,
For 't was the seat of Persian kings renowned.
Here stood the royal houses of delight
Where kings have shown their glory, wealth, and might,
The sumptuous palace of Queen Esther here,
And of good Mordecai, her kinsman dear.
Those purple hangings mixed with green and white,
Those beds of gold and couches of delight,
And furniture the richest in all lands
Now fall into the Macedonian's hands.
From Shushan to Persepolis he goes,
Which news doth still augment Darius' woes.
In his approach the governor sends word
For his receipt with joy they all accord:
With open gates the wealthy town did stand,
And all in it was at his high command.
Of all the cities that on earth were found
None like to this in riches did abound.
Though Babylon was rich, and Shushan, too,
Yet to compare with this they might not do.
Here lay the bulk of all those precious things
That did pertain unto the Persian kings,
For when the soldiers rifled had their pleasure,
And taken money, plate, and golden treasure,

The Four Monarchies

Statues, some gold, and silver numberless,
Yet after all, as stories do express,
The share of Alexander did amount
To an hundred thousand talents by account.
Here of his own he sets a garrison,
As first at Shushan and at Babylon.
On their old governors titles he laid,
But on their faithfulness he never stayed —
Their places gave to his captains, as was just,
For such revolters false what king can trust?
The riches and the pleasures of this town
Now make this king his virtues all to drown.
He walloweth in all licentiousness,
In pride and cruelty to high excess.
Being inflamed with wine upon a season,
Filléd with madness, and quite void of reason,
He at a bold proud strumpet's lewd desire
Commands to set this goodly town on fire.
Parmenio wise entreats him to desist,
And lays before his eyes, if he persist,
His fame's dishonor, loss unto his state,
And just procuring of the Persians' hate.
But, deaf to reason, bent to have his will,
Those stately streets with raging flame did fill.
Then to Darius he directs his way,
Who was retired as far as Media,
And there, with sorrows, fears, and cares surrounded,
Had now his army fourth and last compounded,

Which forty thousand made. Now his intent
Was these in Bactria soon to augment;
But, hearing Alexander was so near,
Thought now this once to try his fortunes here,
And rather choose an honorable death
Than still with infamy to draw his breath.
But Bessus false, who was his chief commander,
Persuades him not to fight with Alexander.
With sage advice he sets before his eyes
The little hope of profit like to rise;
If when he'd multitudes the day he lost,
Then with so few how likely to be crossed.
This counsel for his safety he pretended,
But to deliver him to his foe intended.
Next day this treason to Darius known,
Transported sore with grief and passiön,
Grinding his teeth, and plucking off his hair,
He sat o'erwhelmed with sorrow and despair;
Then bids his servant Artabazus true
Look to himself, and leave him to that crew
Who was of hopes and comforts quite bereft
And by his guard and servitors all left.
Straight Bessus comes, and with his traitorous hands
Lays hold on his lord, and, binding him with bands,
Throws him into a cart covered with hides,—
Who, wanting means to resist these wrongs, abides,—
Then draws the cart along with chains of gold
In more despite the thrallèd prince to hold.

And thus toward Alexander on he goes.
Great recompense for this he did propose.
But some, detesting this his wicked fact,
To Alexander flies and tells this act,
Who, doubling of his march, posts on amain
Darius from that traitor's hands to gain.
Bessus gets knowledge his disloyalty
Had Alexander's wrath incensèd high,
Whose army now was almost within sight.
His hopes being dashed, prepares himself for flight.
Unto Darius first he brings a horse,
And bids him save himself by speedy course.
The woeful king his courtesy refuses,
Whom thus the execrable wretch abuses:
By throwing darts gave him his mortal wound,
Then slew his servants that were faithful found,
Yea, wounds the beasts that drew him unto death,
And leaves him thus to gasp out his last breath.
Bessus his partner in this tragedy
Was the false governor of Media.
This done, they with their host soon speed away
To hide themselves remote in Bactria.
Darius, bathed in blood, sends out his groans,
Invokes the heavens and earth to hear his moans;
His lost felicity did grieve him sore,
But this unheard-of treachery much more —
But above all that neither ear nor eye
Should hear nor see his dying misery.

As thus he lay, Polystratus, a Greek,
Wearied with his long march, did water seek,
So chanced these bloody horses to espy,
Whose wounds had made their skins of purple dye;
To them repairs, then, looking in the cart,
Finds poor Darius piercéd to the heart,
Who, not a little cheered to have some eye
The witness of this horrid tragedy,
Prays him to Alexander to commend
The just revenge of this his woeful end,
And not to pardon such disloyalty
Of treason, murder, and base cruelty —
If not because Darius thus did pray,
Yet that succeeding kings in safety may
Their lives enjoy, their crowns, and dignity,
And not by traitors' hands untimely die.
He also sends his humble thankfulness
For all the kingly grace he did express
To his mother, children dear, and wife now gone,
Which made their long restraint seem to be none;
Praying the immortal gods that sea and land
Might be subjected to his royal hand,
And that his rule as far extended be
As men the rising, setting, sun shall see.
This said, the Greek for water doth entreat
To quench his thirst and to allay his heat.
"Of all good things," quoth he, "once in my power,
I've nothing left, at this my dying hour,

Thy service and compassion to reward;
But Alexander will, for this regard."
This said, his fainting breath did fleet away,
And though a monarch late, now lies like clay.
And thus must every son of Adam lie;
Though gods on earth, like sons of men they die.
Now to the East great Alexander goes
To see if any dare his might oppose;
For scarce the world or any bounds thereon
Could bound his boundless fond ambitiön.
Such as submit, again he doth restore
Their riches, and their honors he makes more;
On Artabazus more than all bestowed
For his fidelity to his master showed.
Thalestris, queen of the Amazons, now brought
Her train to Alexander, as 't is thought;
Though most of reading best and soundest mind
Such country there nor yet such people find.
Than tell her errand we had better spare;
To the ignorant her title will declare.
As Alexander in his greatness grows,
So daily of his virtues doth he lose:
He baseness counts his former clemency,
And not beseeming such a dignity;
His past sobriety doth also 'bate,
As most incompatible to his state;
His temperance is but a sordid thing,
Noways becoming such a mighty king.

His greatness now he takes to represent
His fancied gods above the firmament,
And such as showed but reverence before
Now are commanded strictly to adore.
With Persian robes himself doth dignify,
Charging the same on his nobility;
His manners, habits, gestures, all did fashion
After that conquered and luxurious nation.
His captains that were virtuously inclined
Grieved at this change of manners and of mind;
The ruder sort did openly deride
His feignéd deity and foolish pride.
The certainty of both comes to his ears,
But yet no notice takes of what he hears.
With those of worth he still desires esteem,
So heaps up gifts his credit to redeem,
And for the rest new wars and travels finds
That other matters might take up their minds.
And hearing Bessus makes himself a king,
Intends that traitor to his end to bring.
Now that his host from luggage might be free,
And with his burden no man burdened be,
Commands forthwith each man his fardel bring
Into the market-place before the king;
Which done, sets fire upon those goodly spoils,
The recompense of travels, wars, and toils,
And thus unwisely in a madding fume
The wealth of many kingdoms did consume.

But marvel 't is that without mutiny
The soldiers should let pass this injury;
Nor wonder less to readers may it bring
Here to observe the rashness of the king.
Now with his army doth he post away
False Bessus to find out in Bactria;
But, much distressed for water in their march,
The drought and heat their bodies sore did parch.
At length they came to the river Oxus' brink,
Where most immoderately these thirsty drink,
Which more mortality to them did bring
Than all their wars against the Persian king.
Here Alexander 's almost at a stand
To pass the river to the other land;
Boats here are none, nor near it any wood
To make them rafts to waft them o'er the flood.
But he that was resolvéd in his mind
Would without means some transportation find.
Then from the carriages the hides he takes,
And, stuffing them with straw, he bundles makes;
On these together tied in six days' space
They all pass over to the other place.
Had Bessus there but valor to his will,
With little pain he might have kept them still.
Coward, he durst not fight, nor could he fly;
Hated of all for his former treachery,
He 's by his own now bound in iron chains,
A collar of the same his neck contains,

And in this sort they rather drag than bring
This malefactor vile before the king,
Who to Darius' brother gives the wretch
With racks and tortures every limb to stretch.
Here was of Greeks a town in Bactria
Whom Xerxes from their country led away.
These not a little joyed this day to see
Wherein their own had got the sovereignty,
And, now revived, with hopes held up their head
From bondage long to be enfranchiséd.
But Alexander puts them to the sword
Without least cause from them in deed or word,
Nor sex nor age, nor one nor other spared,
But in his cruelty alike they shared;
Nor reason could he give for this great wrong
But that they had forgot their mother tongue.
While thus some time he spent in Bactria,
And in his camp strong and securely lay,
Down from the mountains twenty thousand came
And there most fiercely set upon the same;
Repelling these, two marks of honor got,
Imprinted in his leg by arrows shot.
The Bactrians against him now rebel;
But he their stubbornness in time doth quell.
From thence he to Jaxartes river goes,
Where Scythians rude his army do oppose,
And with their outcries in an hideous sort
Beset his camp, or military court,

Of darts and arrows made so little spare
They flew so thick they seemed to dark the air.
But soon his soldiers forced them to a flight;
Their nakedness could not endure their might.
Upon this river's bank in seventeen days
A goodly city doth completely raise,
Which Alexandria he doth likewise name,
And sixty furlongs could but round the same.
A third supply Antipater now sent,
Which did his former forces much augment;
And being one hundred twenty thousand strong,
He enters then the Indian kings among.
Those that submit, he gives them rule again;
Such as do not, both they and theirs are slain.
His wars with sundry nations I'll omit,
And also of the Malliens what is writ:
His fights, his dangers, and the hurts he had;
How to submit their necks at last they're glad.
To Nysa he goes, by Bacchus built long since,
Whose feasts are celebrated by this prince;
Nor had that drunken god one who would take
His liquors more devoutly for his sake.
When thus ten days his brain with wine he'd soaked,
And with delicious meats his palate choked,
To the river Indus next his course he bends.
Boats to prepare Hephæstion first he sends,
Who, coming thither long before his lord,
Had to his mind made all things to accord;

The vessels ready were at his command,
And Omphis, king of that part of the land,
Through his persuasion Alexander meets
And as his sovereign lord him humbly greets.
Fifty-six elephants he brings to his hand,
And tenders him the strength of all his land;
Presents himself first with a golden crown,
Then eighty talents to his captains down.
But Alexander made him to behold
He glory sought, no silver, no, nor gold;
His presents all with thanks he did restore,
And of his own a thousand talents more.
Thus all the Indian kings to him submit
But Porus stout, who will not yield as yet.
To him doth Alexander thus declare:
His pleasure is that forthwith he repair
Unto his kingdom's borders and, as due,
His homage to himself as sovereign do.
But kingly Porus this brave answer sent,
That to attend him there was his intent,
And come as well provided as he could;
But for the rest his sword advise him should.
Great Alexander, vexed at this reply,
Did more his valor than his crown envy;
He's now resolved to pass Hydaspes' flood,
And there by force his sovereignty make good.
Stout Porus on the banks doth ready stand
To give him welcome when he comes to land;

The Four Monarchies

A potent army with him, like a king,
And ninety elephants for war did bring.
Had Alexander such resistance seen
On Tigris' side, here now he had not been.
Within this spacious river, deep and wide,
Did here and there isles full of trees abide.
His army Alexander doth divide,
With Ptolemy sends part to the other side;
Porus encounters them, thinks all are there,
When covertly the rest get o'er elsewhere,
And whilst the first he valiantly assailed,
The last set on his back, and so prevailed.
Yet work enough here Alexander found,
For to the last stout Porus kept his ground;
Nor was it dishonor at length to yield
When Alexander strives to win the field.
The kingly captive 'fore the victor 's brought;
In looks or gesture not abaséd aught,
But him a prince of an undaunted mind
Did Alexander by his answers find.
His fortitude his royal foe commends,
Restores him, and his bounds farther extends.
Now eastward Alexander would go still,
But so to do his soldiers had no will;
Long with excessive travels weariéd,
Could by no means be farther drawn or led.
Yet that his fame might to posterity
Be had in everlasting memory,

He for his camp a greater circuit takes,
And for his soldiers larger cabins makes;
His mangers he erected up so high
As never horse his provender could eye;
Huge bridles made, which here and there he left,
Which might be found and for great wonders kept.
Twelve altars then for monuments he rears,
Whereon his acts and travels long appear;
But doubting wearing time might these decay,
And so his memory would fade away,
He on the fair Hydaspes' pleasant side
Two cities built, his name might there abide:
First, Nicæa; the next, Bucephalon,
Where he entombed his stately stalliön.
His fourth and last supply was hither sent,
Then down Hydaspes with his fleet he went.
Some time he after spent upon that shore,
Whither ambassadors, ninety or more,
Came with submission from the Indian kings,
Bringing their presents rare and precious things.
These all he feasts in state on beds of gold,
His furniture most sumptuous to behold;
His meat and drink, attendants, everything,
To the utmost showed the glory of a king.
With rich rewards he sent them home again,
Acknowledging their masters sovereign.
Then sailing south and coming to that shore,
Those obscure nations yielded as before.

A city here he built, called by his name,
Which could not sound too oft with too much fame.
Then sailing by the mouth of Indus' flood,
His galleys stuck upon the flats and mud,
Which the stout Macedonians amazéd sore,
Deprived at once the use of sail and oar.
Observing well the nature of the tide,
In those their fears they did not long abide.
Passing fair Indus' mouth, his course he steered
To the coast which by Euphrates' mouth appeared,
Whose inlets near unto he winter spent,
Unto his starvéd soldiers' small content —
By hunger and by cold so many slain
That of them all the fourth did scarce remain.
Thus winter, soldiers, and provisions spent,
From thence he then unto Gedrosia went;
And thence he marched into Carmania,
And so at length drew near to Persiä.
Now through these goodly countries as he passed
Much time in feasts and rioting did waste.
Then visits Cyrus' sepulcher in his way,
Who now obscure at Pasargadæ lay;
Upon his monument his robe he spread,
And set his crown on his supposéd head.
From thence to Babylon; some time there spent.
He at the last to royal Shushan went.
A wedding feast to his nobles then he makes,
And Statira, Darius' daughter, takes;

Her sister gives to his Hephæstion dear,
That by this match he might be yet more near.
He fourscore Persian ladies also gave
At this same time unto his captains brave.
Six thousand guests unto this feast invites,
Whose senses all were glutted with delights.
It far exceeds my mean abilities
To shadow forth these short felicities;
Spectators here could scarce relate the story,
They were so rapt with this external glory.
If an ideal paradise a man would frame,
He might this feast imagine by the same.
To every guest a cup of gold he sends.
So after many days the banquet ends.
Now Alexander's conquests all are done,
And his long travails past and overgone;
His virtues dead, buried, and quite forgot,
But vice remains to his eternal blot.
'Mongst those that of his cruelty did taste
Philotas was not least nor yet the last.
Accused because he did not certify
The king of treason and conspiracy,
Upon suspicion being apprehended,
Nothing was proved wherein he had offended
But silence, which was of such consequence
He was judged guilty of the same offense.
But for his father's great deserts the king
His royal pardon gave for this foul thing.

Yet is Philotas unto judgment brought,
Must suffer, not for what is proved, but thought.
His master is accuser, judge, and king,
Who to the height doth aggravate each thing,
Inveighs against his father now absent,
And his brethren who for him their lives had spent.
But Philotas his unpardonable crime
No merit could obliterate or time:
He did the oracle of Jove deride
By which His Majesty was deified.
Philotas, thus o'ercharged with wrong and grief,
Sunk in despair without hope of relief,
Fain would have spoke and made his own defense;
The king would give no ear, but went from thence,
To his malicious foes delivers him
To wreak their spite and hate on every limb.
Philotas after him sends out this cry:
"O Alexander, thy free clemency
My foes exceeds in malice, and their hate
Thy kingly word can easily terminate."
Such torments great as wit could worst invent
Or flesh and life could bear till both were spent
Were now inflicted on Parmenio's son,
He might accuse himself, as they had done;
At last he did, so they were justified,
And told the world that for his guilt he died.
But how these captains should, or yet their master,
Look on Parmenio after this disaster

They knew not, wherefore best now to be done
Was to despatch the father as the son.
This sound advice at heart pleased Alexander,
Who was so much engaged to this commander
As he would ne'er confess nor yet reward,
Nor could his captains bear so great regard;
Wherefore at once, all these to satisfy,
It was decreed Parmenio should die.
Polydamas, who seemed Parmenio's friend,
To do this deed they into Media send;
He walking in his garden to and fro,
Fearing no harm, because he none did do,
Most wickedly was slain without least crime,
The most renownéd captain of his time.
This is Parmenio who so much had done
For Philip dead and his surviving son,
Who from a petty king of Macedon
By him was set upon the Persian throne;
This that Parmenio who still overcame,
Yet gave his master the immortal fame;
Who for his prudence, valor, care, and trust
Had this reward, most cruel and unjust.
The next who in untimely death had part
Was one of more esteem but less desert —
Clitus, beloved next to Hephæstiön,
And in his cups his chief companiön.
When both were drunk, Clitus was wont to jeer;
Alexander to rage, to kill, and swear.

The Four Monarchies

Nothing more pleasing to mad Clitus' tongue
Than his master's godhead to defy and wrong;
Nothing touched Alexander to the quick
Like this against his deity to kick.
Both at a feast, when they had tippled well,
Upon this dangerous theme fond Clitus fell;
From jest to earnest, and at last so bold
That of Parmenio's death him plainly told,
Which Alexander's wrath incensed so high
Naught but his life for this could satisfy.
From one stood by he snatched a partizan,
And in a rage him through the body ran.
Next day he tore his face for what he'd done,
And would have slain himself for Clitus gone;
This pot companion he did more bemoan
Than all the wrongs to brave Parmenio done.
The next of worth that suffered after these
Was learnéd, virtuous, wise Callisthenes,
Who loved his master more than did the rest,
As did appear in flattering him the least.
In his esteem a god he could not be,
Nor would adore him for a deity;
For this alone, and for no other cause
Against his sovereign or against his laws,
He on the rack his limbs in pieces rent.
Thus was he tortured till his life was spent.
On this unkingly act doth Seneca
This censure pass, and not unwisely say

Of Alexander this the eternal crime,
Which shall not be obliterate by time,
Which virtue's fame can ne'er redeem by far,
Nor all felicity of his in war.
Whene'er 't is said he thousand thousands slew,—
Yea, and Callisthenes to death he drew.
The mighty Persian king he overcame,—
Yea, and he killed Callisthenes of fame.
All countries, kingdoms, provinces, he won
From Hellespont to the farthest oceän;
All this he did, who knows not to be true?—
But yet, withal, Callisthenes he slew.
From Macedon his empire did extend
Unto the utmost bounds of the orient;
All this he did, yea, and much more, 't is true,—
But yet, withal, Callisthenes he slew.
Now Alexander goes to Media,
Finds there the want of wise Parmenio.
Here his chief favorite, Hephæstion, dies.
He celebrates his mournful obsequies;
Hangs his physiciän—the reason why,
He sufferéd his friend Hephæstion die.
This act, methinks, his godhead should ashame,
To punish where himself deservéd blame;
Or of necessity he must imply
The other was the greatest deity.
The mules and horses are for sorrow shorn.
The battlements from off the walls are torn

The Four Monarchies

Of stately Ecbatana, who now must show
A rueful face in this so general woe.
Twelve thousand talents also did intend
Upon a sumptuous monument to spend.
Whate'er he did or thought, not so content,
His messenger to Jupiter he sent,
That by his leave his friend Hephæstiön
Among the demigods they might enthrone.
From Media to Babylon he went;
To meet him there to Antipater he'd sent,
That he might act also upon the stage
And in a tragedy there end his age.
The queen Olympias bears him deadly hate,
Not suffering her to meddle with the state,
And by her letters did her son incite
This great indignity he should requite.
His doing so no whit displeased the king,
Though to his mother he disproved the thing.
But now Antipater had lived so long
He might well die, though he had done no wrong;
His service great is suddenly forgot,
Or, if remembered, yet regarded not.
The king doth intimate 't was his intent
His honors and his riches to augment,
Of larger provinces the rule to give,
And for his counsel near the king to live.
So to be caught Antipater's too wise;
Parmenio's death's too fresh before his eyes.

He was too subtile for his crafty foe,
Nor by his baits could be ensnaréd so;
But his excuse with humble thanks he sends,
His age and journey long he then pretends,
And pardon craves for his unwilling stay;
He shows his grief he 's forced to disobey.
Before his answer came to Babylon
The thread of Alexander's life was spun;
Poison had put an end to his days, 't was thought,
By Philip and Cassander to him brought,
Sons to Antipater, and bearers of his cup,
Lest of such like their father chance to sup.
But others thought, and that more generally,
That through excessive drinking he did die.
The thirty-third of his age do all agree
This conqueror did yield to destiny.
When this sad news came to Darius' mother,
She laid it more to heart than any other,
Nor meat, nor drink, nor comfort would she take,
But pined in grief till life did her forsake;
All friends she shunned, yea, banishéd the light,
Till death enwrapped her in perpetual night.
This monarch's fame must last whilst world doth stand,
And conquests be talked of whilst there is land;
His princely qualities had he retained,
Unparalleled for ever had remained.
But with the world his virtues overcame,
And so with black beclouded all his fame.

Wise Aristotle, tutor to his youth,
Had so instructed him in moral truth,
The principles of what he then had learned
Might to the last, when sober, be discerned.
Learning and learnéd men he much regarded,
And curious artists evermore rewarded.
The Iliad of Homer he still kept,
And under his pillow laid it when he slept.
Achilles' happiness he did envy,
'Cause Homer kept his acts to memory.
Profusely bountiful without desert,
For such as pleased him had both wealth and heart;
Cruel by nature and by custom, too,
As oft his acts throughout his reign do show;
Ambitious so that naught could satisfy,
Vain, thirsting after immortality,
Still fearing that his name might hap to die,
And fame not last unto eternity.
This conqueror did oft lament, 't is said,
There were no more worlds to be conqueréd.
This folly great Augustus did deride,
For had he had but wisdom to his pride
He would have found enough there to be done
To govern that he had already won.
His thoughts are perished, he aspires no more,
Nor can he kill or save as heretofore.
A god alive, him all must idolize;
Now like a mortal, helpless man he lies.

Of all those kingdoms large which he had got
To his posterity remained no jot;
For by that hand which still revengeth blood
None of his kindred nor his race long stood;
But as he took delight much blood to spill,
So the same cup to his did others fill.
Four of his captains now do all divide,
As Daniël before had prophesied.
The leopard down, the four wings 'gan to rise,
The great horn broke, the less did tyrannize.
What troubles and contentions did ensue
We may hereafter show in season due.

ARRHIDÆUS.

Great Alexander dead, his army's left
Like to that giant of his eye bereft.
(When of his monstrous bulk it was the guide,
His matchless force no creature could abide;
But by Ulysses having lost his sight,
All men began straight to contemn his might;
For, aiming still amiss, his dreadful blows
Did harm himself, but never reached his foes.)
Now court and camp all in confusion be.
A king they'll have, but who none can agree;
Each captain wished this prize to bear away,
But none so hardy found as so durst say.
Great Alexander did leave issue none,
Except by Artabazus' daughter one;

The Four Monarchies

And Roxane fair, whom late he markiéd,
Was near her time to be deliveréd.
By nature's right these had enough to claim,
But meanness of their mothers barred the same,
Alleged by those who by their subtile plea
Had hope themselves to bear the crown away.
A sister Alexander had, but she
Claimed not; perhaps her sex might hindrance be.
After much tumult they at last proclaimed
His base-born brother, Arrhidæus named,
That so under his feeble wit and reign
Their ends they might the better still attain.
This choice Perdiccas vehemently disclaimed,
And babe unborn of Roxane he proclaimed.
Some wishéd him to take the style of king,
Because his master gave to him his ring,
And had to him still since Hephæstion died
More than the rest his favor testified;
But he refused, with feignéd modesty,
Hoping to be elect more generally.
He hold on this occasion should have laid,
For second offer there was never made.
'Mongst these contentions, tumults, jealousies,
Seven days the corpse of their great master lies
Untouched, uncovered, slighted, and neglected,
So much these princes their own ends respected—
A contemplation to astonish kings,
That he who late possessed all earthly things,

And yet not so content unless that he
Might be esteeméd for a deity,
Now lay a spectacle to testify
The wretchedness of man's mortality.
After some time, when stirs began to calm,
His body did the Egyptians embalm;
His countenance so lively did appear
That for a while they durst not come so near.
No sign of poison in his entrails found,
But all his bowels colored well and sound.
Perdiccas, seeing Arrhidæus must be king,
Under his name began to rule each thing.
His chief opponent who controlled his sway
Was Meleager, whom he would take away;
And by a wile he got him in his power,
So took his life unworthily that hour,
Using the name and the command of the king
To authorize his acts in everything.
The princes, seeing Perdiccas' power and pride,
For their security did now provide.
Antigonus for his share Asia takes,
And Ptolemy next sure of Egypt makes;
Seleucus afterward held Babylon;
Antipater had long ruled Macedon.
These now to govern for the king pretend,
But nothing less each one himself intends.
Perdiccas took no province like the rest,
But held command of the army, which was best,

The Four Monarchies

And had a higher project in his head —
His master's sister secretly to wed.
So to the lady covertly he sent,
That none might know to frustrate his intent.
But Cleopatra this suitor did deny
For Leonnatus, more lovely in her eye,
To whom she sent a message of her mind
That if he came good welcome he should find.
In these tumultuous days the thrallèd Greeks
Their ancient liberty afresh now seek,
And gladly would the yoke shake off laid on
Sometime by Philip and his conquering son.
The Athenians force Antipater to fly
To Lamia, where he shut up doth lie.
To brave Craterus then he sends with speed
For succor to relieve him in his need.
The like of Leonnatus he requires —
Which at this time well suited his desires,
For to Antipater he now might go,
His lady take in the way, and no man know.
Antiphilus, the Athenian general,
With speed his army doth together call
And Leonnatus seeks to stop, that so
He join not with Antipater their foe.
The Athenian army was the greater far,
Which did his match with Cleopatra mar,
For, fighting still while there did hope remain,
The valiant chief amidst his foes was slain.

'Mongst all the princes of great Alexander
For person none was like to this commander.
Now to Antipater Craterus goes,
Blocked up in Lamia still by his foes;
Long marches through Ciliciā he makes,
And the remains of Leonnatus takes.
With them and his he into Grecia went,
Antipater released from prisonment.
After which time the Greeks did nevermore
Act anything of worth as heretofore,
But under servitude their necks remained,
Nor former liberty or glory gained.
Now died, about the end of the Lamian war,
Demosthenes, that sweet-tongued orator,
Who feared Antipater would take his life
For animating the Athenian strife,
And to end his days by poison rather chose
Than fall into the hands of mortal foes.
Craterus and Antipater now join,
In love and in affinity combine:
Craterus doth his daughter Phila wed
Their friendship might the more be strengthenéd.
Whilst they in Macedon do thus agree,
In Asiā they all asunder be.
Perdiccas grieved to see the princes bold
So many kingdoms in their power to hold,
Yet to regain them how he did not know;
His soldiers 'gainst those captains would not go.

To suffer them go on as they begun
Was to give way himself might be undone.
With Antipater to join he sometimes thought,
That by his help the rest might low be brought,
But this again dislikes; he would remain,
If not in style, in deed a sovereign —
For all the princes of great Alexander
Acknowledgéd for chief that old commander.
Desires the king to go to Macedon,
Which once was of his ancestors the throne,
And by his presence there to nullify
The acts of his viceroy now grown so high.
Antigonus of treason first attaints,
And summons him to answer his complaints.
This he avoids, and ships himself and son,
Goes to Antipater and tells what's done.
He and Craterus both with him do join,
And 'gainst Perdiccas all their strength combine.
Brave Ptolemy to make a fourth then sent,
To save himself from danger imminent;
In midst of these garboils, with wondrous state
His master's funeral doth celebrate;
In Alexandria his tomb he placed,
Which eating time hath scarcely yet defaced.
Two years and more since nature's debt he paid,
And yet till now at quiet was not laid.
Great love did Ptolemy by this act gain,
And made the soldiers on his side remain.

Perdiccas hears his foes are all combined.
'Gainst which to go he 's not resolved in mind,
But first 'gainst Ptolemy he judged was best,—
Nearest to him, and farthest from the rest,—
Leaves Eumenes the Asian coast to free
From the invasions of the other three,
And with his army unto Egypt goes
Brave Ptolemy to the utmost to oppose.
Perdiccas' surly carriage and his pride
Did alienate the soldiers from his side;
But Ptolemy, by his affability,
His sweet demeanor, and his courtesy,
Did make his own firm to his cause remain,
And from the other side did daily gain.
Perdiccas in his pride did ill entreat
Pithon, of haughty mind and courage great,
Who could not brook so great indignity,
But of his wrongs his friends doth certify;
The soldiers 'gainst Perdiccas they incense,
Who vow to make this captain recompense,
And in a rage they, rushing to his tent,
Knock out his brains; to Ptolemy then went
And offer him his honors and his place,
With style of the Protector him to grace.
Next day into the camp came Ptolemy,
And is received of all most joyfully.
Their proffers he refused with modesty,
Yields them to Pithon for his courtesy.

The Four Monarchies

With what he held he was now more content
Than by more trouble to grow eminent.
Now comes the news of a great victory
That Eumenes got of the other three.
Had it but in Perdiccas' life arrived
With greater joy it would have been received.
Thus Ptolemy rich Egypt did retain,
And Pithon turned to Asiä again.
Whilst Perdiccas encamped in Africa,
Antigonus did enter Asiä,
And fain would Eumenes draw to their side;
But he alone most faithful did abide.
The others all had kingdoms in their eye,
But he was true to his master's family.
Nor could Craterus, whom he much did love,
From his fidelity once make him move.
Two battles he fought, and had of both the best,
And brave Craterus slew among the rest;
For this sad strife he pours out his complaints,
And his beloved foe full sore laments.
I should but snip a story into bits,
And his great acts and glory much eclipse,
To show the dangers Eumenes befel,
His stratagems wherein he did excel,
His policies, how he did extricate
Himself from out of labyrinths intricate.
He that at large would satisfy his mind
In Plutarch's Lives his history may find.

For all that should be said let this suffice,
He was both valiant, faithful, patient, wise.
Pithon's now chosen protector of the state;
His rule Queen Eurydice begins to hate,
Sees Arrhidæus must not king it long
If once young Alexander grow more strong.
But that her husband serve for supplement
To warm his seat was never her intent.
She knew her birthright gave her Macedon,
Grandchild to him who once sat on that throne,
Who was Perdiccas, Philip's eldest brother,
She daughter to his son, who had no other.
Pithon commands, as oft she countermands;
What he appoints, she purposely withstands.
He, wearied out at last, would needs be gone,
Resigned his place, and so let all alone.
In his room the soldiers chose Antipater,
Who vexed the queen more than the other far.
From Macedon to Asiä he came
That he might settle matters in the same.
He placed, displaced, controlled, ruled, as he list,
And this no man durst question or resist;
For all the nobles of King Alexander
Their bonnets veiled to him as chief commander.
When to his pleasure all things they had done,
The king and queen he takes to Macedon,
Two sons of Alexander, and the rest,
All to be ordered there as he thought best.

The army to Antigonus doth leave,
And government of Asia to him gave;
And thus Antipater the groundwork lays
On which Antigonus his height doth raise,
Who in few years the rest so overtops
For universal monarchy he hopes.
With Eumenes he divers battles fought,
And by his sleights to circumvent him sought;
But vain it was to use his policy
'Gainst him that all deceits could scan and try.
In this epitome too long to tell
How finely Eumenes did here excel,
And by the self-same traps the other laid
He to his cost was righteously repaid.
But while these chieftains do in Asia fight,
To Greece and Macedon let's turn our sight.
When great Antipater the world must leave,
His place to Polysperchon he did bequeath,
Fearing his son Cassander was unstaid,
Too rash to bear that charge, if on him laid.
Antigonus, hearing of his decease,
On most part of Assyria doth seize.
And Ptolemy next to encroach begins;
All Syria and Pheniciā he wins.
Then Polysperchon begins to act in his place,
Recalls Olympias the court to grace.
Antipater had banished her from thence
Into Epirus for her great turbulence;

This new protector's of another mind,
Thinks by her majesty much help to find.
Cassander like his father could not see
This Polysperchon's great ability,
Slights his commands, his actions he disclaims,
And to be chief himself now bends his aims.
Such as his father had advanced to place,
Or by his favors any way had graced,
Are now at the devotion of the son,
Pressed to accomplish what he would have done.
Besides, he was the young queen's favorite,
On whom, 't was thought, she set her chief delight.
Unto these helps at home he seeks out more,
Goes to Antigonus and doth implore,
By all the bonds 'twixt him and his father past,
And for that great gift which he gave him last,
By these and all to grant him some supply
To take down Polysperchon grown so high.
For this Antigonus did need no spurs,
Hoping to gain yet more by these new stirs,
Straight furnished him with a sufficient aid,
And so he quick returns thus well appaid;
With ships at sea, an army for the land,
His proud opponent he hopes soon to withstand.
But in his absence Polysperchon takes
Such friends away as for his interest makes
By death, by prison, or by banishment,
That no supply by these here might be lent.

Cassander with his host to Grecia goes,
Whom Polysperchon labors to oppose,
But beaten was at sea and foiled at land.
Cassander's forces had the upper hand.
Athens with many towns in Greece beside
Firm for his father's sake to him abide.
Whilst hot in wars these two in Greece remain,
Antigonus doth all in Asia gain;
Still labors Eumenes would with him side,
But all in vain; he faithful did abide,
Nor mother could nor sons of Alexander
Put trust in any but in this commander.
The great ones now begin to show their mind,
And act as opportunity they find.
Arrhidæus, the scorned and simple king,
More than he bidden was could act no thing.
Polysperchon, for office hoping long,
Thinks to enthrone the prince when riper grown.
Eurydice this injury disdains,
And to Cassander of this wrong complains.
Hateful the name and house of Alexander
Was to this proud and vindictive Cassander;
He still kept locked within his memory
His father's danger, with his family,
Nor thought he that indignity was small
When Alexander knocked his head to the wall.
These, with his love unto the amorous queen,
Did make him vow her servant to be seen.

Olympias Arrhidæus deadly hates,
As all her husband's children by his mates;
She gave him poison formerly, 't is thought,
Which damage both to mind and body brought.
She now with Polysperchon doth combine
To make the king by force his seat resign,
And her young grandchild in his state enthrone,
That under him she might rule all alone.
For aid she goes to Epirus among her friends
The better to accomplish these her ends.
Eurydice, hearing what she intends,
In haste unto her friend Cassander sends
To leave his siege at Tegea, and with speed
To save the king and her in this their need;
Then by entreaties, promises, and coin
Some forces did procure with her to join.
Olympias soon enters Macedon.
The queen to meet her bravely marches on;
But when her soldiers saw their ancient queen,
Calling to mind what sometime she had been,—
The wife and mother of their famous kings,—
Nor darts nor arrows now none shoots or flings.
The king and queen, seeing their destiny,
To save their lives to Amphipolis do fly;
But the old queen pursues them with her hate,
And needs will have their lives as well as state.
The king by extreme torments had his end,
And to the queen these presents she did send—

The Four Monarchies

A halter, cup of poison, and a sword,
Bids choose her death, such kindness she 'll afford.
The queen, with many a curse and bitter check,
At length yields to the halter her fair neck,
Praying that fatal day might quickly haste
On which Olympias of the like might taste.
This done, the cruel queen rests not content.
'Gainst all that loved Cassander she was bent:
His brethren, kinsfolk, and his chiefest friends
That fell within her reach came to their ends;
Digged up his brother dead, 'gainst nature's right,
And threw his bones about to show her spite.
The courtiers, wondering at her furious mind,
Wished in Epirus she 'd been still confined.
In Peloponnesus then Cassander lay,
Where hearing of this news he speeds away;
With rage and with revenge he 's hurried on
To find this cruel queen in Macedon.
But being stopped at strait Thermopylæ,
Sea-passage gets, and lands in Thessaly;
His army he divides, sends part away
Polysperchon to hold a while in play,
And with the rest Olympias pursues
For all her cruelty to give her dues.
She with the chief of court to Pydna flies;
Well fortified and on the sea it lies;
There by Cassander she 's blocked up so long
Until the famine grows exceeding strong.

Her cousin of Epirus did what he might
To raise the siege and put her foes to flight;
Cassander is resolved there to remain,
So succors and endeavors prove but vain.
Fain would this wretched queen capitulate;
Her foe would give no ear, such is his hate.
The soldiers, pinchéd with this scarcity,
By stealth unto Cassander daily fly.
Olympias means to hold out to the last,
Expecting nothing but of death to taste;
But his occasions calling him away,
Gives promise for her life, so wins the day.
No sooner had he got her in his hand
But he made in judgment her accusers stand
And plead the blood of friends and kindred spilt,
Desiring justice might be done for guilt;
And so was he acquitted of his word,
For justice' sake she being put to the sword.
This was the end of this most cruel queen,
Whose fury scarcely paralleled hath been —
The daughter, sister, mother, wife, to kings;
But royalty no good conditions brings.
To her husband's death,'t is thought, she gave consent,
The murderer she did so much lament,
With garlands crowned his head, bemoaned his fate,
His sword did to Apollo consecrate.
Her outrages too tedious are to relate —
How for no cause but her inveterate hate

The Four Monarchies

Her husband's wives and children after his death
Some slew, some fried, of others stopped the breath.
Now in her age she's forced to taste that cup
Which she had others often made to sup.
Now many towns in Macedon suppressed,
And Pella's fain to yield among the rest.
The funerals Cassander celebrates
Of Arrhidæus and his queen with state;
Among their ancestors by him they're laid,
And shows of lamentation for them made.
Old Thebes he then rebuilt, so much of fame,
And Cassandria raised after his name.
But leave him building, others in their urn;
Let's for a while now into Asia turn.
True Eumenes endeavors by all skill
To keep Antigonus from Shushan still;
Having command of the treasure, he can hire
Such as no threats nor favor could acquire.
In divers battles he had good success;
Antigonus came off still honorless.
When victor oft he'd been, and so might still,
Peucestes did betray him by a wile
To Antigonus, who took his life unjust
Because he never would forego his trust.
Thus lost he all for his fidelity,
Striving to uphold his master's family.
But to a period as that did haste,
So Eumenes, the prop, of death must taste.

All Persia now Antigonus doth gain,
And master of the treasure sole remain.
Then with Seleucus straight at odds doth fall,
And he for aid to Ptolemy doth call.
The princes all begin now to envy
Antigonus, he growing up so high;
Fearing his force and what might hap ere long,
Enter into a combination strong:
Seleucus, Ptolemy, Cassander, join,
Lysimachus to make a fourth combines.
Antigonus, desirous of the Greeks,
To make Cassander odious to them seeks,
Sends forth his declarations near and far,
And clears what cause he had to make this war;
Cassander's outrages at large doth tell,
Shows his ambitious practices as well:
The mother of their king to death he 'd put,
His wife and son in prison close had shut;
And aiming now to make himself a king,
And that some title he might seem to bring,
Thessalonica he had newly wed,
Daughter to Philip, their renownéd head;
Had built and called a city by his name,
Which none e'er did but those of royal fame;
And in despite of their two famous kings
Hateful Olynthians to Greece rebrings;
Rebellious Thebes he had reëdified,
Which their late king in dust had damnified.

Requires them therefore to take up their arms
And to requite this traitor for these harms.
Then Ptolemy would gain the Greeks likewise,
And he declares the other's injuries:
First how he held the empire in his hands,
Seleucus driven from government and lands,
The valiant Eumenes unjustly slain,
And lord of royal Shushan did remain;
Therefore requests their help to take him down
Before he wear the universal crown.
These princes at the sea soon had a fight,
Where great Antigonus was put to flight.
His son at Gaza likewise lost the field,
So Syria to Ptolemy did yield.
And Seleucus recovers Babylon;
Still gaining countries eastward he goes on.
Demetrius with Ptolemy did fight,
And, coming unawares, put him to flight;
But bravely sends the prisoners back again,
With all the spoil and booty he had ta'en —
Courteous as noble Ptolemy, or more,
Who at Gaza did the like to him before.
Antigonus did much rejoice his son
With victory his lost repute had won.
At last these princes, tired out with wars,
Sought for a peace, and laid aside their jars.
The terms of their agreement thus express,
That each should hold what now he did possess

Till Alexander unto age was grown,
Who then should be installéd in the throne.
This touched Cassander sore for what he 'd done,
Imprisoning both the mother and the son.
He sees the Greeks now favor their young prince
Whom he in durance held now and long since;
That in few years he must be forced or glad
To render up such kingdoms as he had;
Resolves to quit his fears by one deed done,
So puts to death the mother and her son.
This Roxane for her beauty all commend,
But for one act she did just was her end:
No sooner was great Alexander dead
But she Darius' daughters murderéd —
Both thrown into a well to hide her blot;
Perdiccas was her partner in this plot.
The heavens seemed slow in paying her the same,
But at the last the hand of vengeance came,
And for that double fact which she had done
The life of her must go and of her son.
Perdiccas' had before for his amiss,
But by their hands who thought not once of this.
Cassander's deed the princes do detest,
But 't was in show; in heart it pleased them best.
That he is odious to the world they 're glad,
And now they were free lords of what they had.
When this foul tragedy was past and done,
Polysperchon brings up the other son

Called Hercules, and elder than his brother,
But Olympias would have preferred the other.
The Greeks, touched with the murder done of late,
This orphan prince began to compassionate,
Begin to mutter much 'gainst proud Cassander,
And place their hopes on the heir of Alexander.
Cassander feared what might of this ensue,
So Polysperchon to his counsel drew,
And gives Peloponnesus for his hire,
Who slew the prince according to desire.
Thus was the race and house of Alexander
Extinct by this inhuman wretch Cassander.
Antigonus for all this doth not mourn;
He knows to his profit this at last will turn.
But that some title now he might pretend
To Cleopatra doth for marriage send.
Lysimachus and Ptolemy the same,
And lewd Cassander, too, sticks not for shame.
She then in Lydia at Sardis lay,
Where by embassage all these princes pray.
Choice above all of Ptolemy she makes;
With his ambassador her journey takes.
Antigonus' lieutenant stays her still
Until he further know his master's will.
Antigonus now had a wolf by the ears:
To hold her still or let her go he fears;
Resolves at last the princess should be slain,
So hinders him of her he could not gain.

Her women are appointed for this deed;
They for their great reward no better speed,
For by command they straight were put to death
As vile conspirators that stopped her breath.
And now he hopes he 's ordered all so well
The world must needs believe what he doth tell.
Thus Philip's house was quite extinguishéd,
Except Cassander's wife, who 's yet not dead,
And by their means who thought of nothing less
Than vengeance just against them to express.
Now blood was paid with blood for what was done
By cruel father, mother, cruel son.
Thus may we hear, and fear, and ever say,
That Hand is righteous still which doth repay.
These captains now the style of kings do take,
For to their crowns there 's none can title make.
Demetrius first the royal style assumed,
By his example all the rest presumed.
Antigonus, himself to ingratiate,
Doth promise liberty to Athens' state;
With arms and with provision stores them well,
The better 'gainst Cassander to rebel.
Demetrius thither goes, is entertained
Not like a king, but like some god they feigned;
Most grossly base was their great adulation,
Who incense burnt, and offeréd oblation.
These kings afresh fall to their wars again.
Demetrius of Ptolemy doth gain.

'T would be an endless story to relate
Their several battles and their several fate,
Their fights by sea, their victories by land,
How some when down straight got the upper hand.
Antigonus and Seleucus then fight
Near Ephesus, each bringing all his might,
And he that conqueror shall now remain
The lordship of all Asia shall retain.
This day 'twixt these two kings ends all the strife
For here Antigonus lost rule and life,
Nor to his son did e'er one foot remain
Of those vast kingdoms he did sometime gain.
Demetrius with his troops to Athens flies,
Hopes to find succor in his miseries;
But they, adoring in prosperity,
Now shut their gates in his adversity.
He, sorely grieved at this his desperate state,
Tries foes, sith friends will not compassionate.
His peace he then with old Seleucus makes,
Who his fair daughter Stratonice takes.
Antiochus, Seleucus' dear loved son,
Is for this fresh young lady quite undone;
Falls so extremely sick all feared his life,
Yet durst not say he loved his father's wife.
When his disease the skilled physician found,
His father's mind he wittily did sound,
Who did no sooner understand the same
But willingly resigned the beauteous dame.

Cassander now must die, his race is run,
And leave the ill got kingdoms he had won.
Two sons he left, born of King Philip's daughter,
Who had an end put to their days by slaughter.
Which should succeed at variance they fell;
The mother would the youngest might excel;
The eldest, enraged, did play the viper's part,
And with his sword did run her through the heart.
Rather than Philip's race should longer live,
He whom she gave his life her death shall give.
This by Lysimachus was after slain,
Whose daughter he not long before had ta'en.
Demetrius is called in by the youngest son
Against Lysimachus, who from him won;
But he a kingdom more than his friend did eye,
Seized upon that, and slew him traitorously.
Thus Philip's and Cassander's race both gone,
And so falls out to be extinct in one.
And though Cassander diéd in his bed,
His seed to be extirpt was destinéd;
For blood which was decreed that he should spill
Yet must his children pay for father's ill.
Jehu in killing Ahab's house did well;
Yet be avenged must blood of Jezebel.
Demetrius thus Cassander's kingdoms gains,
And now in Macedon as king he reigns.
Though men and money both he hath at will,
In neither finds content if he sits still.

The Four Monarchies

That Seleucus holds Asia grieves him sore;
Those countries large his father got before.
These to recover musters all his might,
And with his son-in-law will needs go fight;
A mighty navy rigged, an army stout,
With these he hopes to turn the world about,
Leaving Antigonus, his eldest son,
In his long absence to rule Macedon.
Demetrius with so many troubles met
As heaven and earth against him had been set;
Disaster on disaster him pursue,
His story seems a fable more than true.
At last he 's taken and imprisonéd
Within an isle that was with pleasures fed;
Enjoyed whate'er beseemed his royalty,
Only restrainéd of his liberty.
After three years he died, left what he 'd won
In Greece unto Antigonus his son;
For his posterity unto this day
Did ne'er regain one foot in Asiā.
His body Seleucus sends to his son,
Whose obsequies with wondrous pomp were done.
Next died the brave and noble Ptolemy,
Renowned for bounty, valor, clemency;
Rich Egypt left, and what else he had won,
To Philadelphus, his more worthy son.
Of the old heroes now but two remain.
Seleucus and Lysimachus, these twain,

Must needs go try their fortune and their might,
And so Lysimachus was slain in fight.
'T was no small joy unto Seleucus' breast
That now he had outlivéd all the rest;
Possession of Europe he thinks to take,
And so himself the only monarch make.
While with these hopes in Greece he did remain
He was by Ptolemy Ceraunus slain,
The second son of the first Ptolemy,
Who for rebellion unto him did fly.
Seleucus was a father and a friend,
Yet by him had this most unworthy end.
Thus with these kingly captains have we done.
A little now how the succession run:
Antigonus, Seleucus, and Cassander,
With Ptolemy, reigned after Alexander;
Cassander's sons soon after his death were slain,
So three successors only did remain;
Antigonus his kingdoms lost and life
Unto Seleucus, author of that strife;
His son Demetrius all Cassander's gains,
And his posterity the same retains;
Demetrius' son was called Antigonus,
And his again was named Demetrius.
I must let pass those many battles fought
Betwixt those kings and noble Pyrrhus stout
And his son Alexander of Epire,
Whereby immortal honor they acquire.

Demetrius had Philip to his son,
Part of whose kingdoms Titus Quintius won;
Philip had Perseus, who was made a thrall
To Æmilius, the Roman general —
Him with his sons in triumph lead did he,
Such riches, too, as Rome did never see.
This of Antigonus his seed 's the fate,
Whose empire was subdued to the Roman state.
Longer Seleucus held the royalty
In Syria by his posterity.
Antiochus Soter his son was named,
To whom the old Berosus, so much famed,
His book of Assur's monarchs dedicates,
Tells of their names, their wars, their riches, fates;
But this is perishéd with many more,
Which oft we wish was extant as before.
Antiochus Theos was Soter's son,
Who a long war with Egypt's king begun;
The affinities and wars Daniel sets forth,
And calls them there the kings of south and north.
This Theos murdered was by his lewd wife.
Seleucus reigned when he had lost his life.
A third Seleucus next sits on the seat,
And then Antiochus surnamed the Great,
Whose large dominions after were made small
By Scipio, the Roman general.
Fourth Seleucus Antiochus succeeds,
And next Epiphanes, whose wicked deeds,

Horrid massacres, murders, cruelties,
Amongst the Jews we read in Maccabees.
Antiochus Eupator was the next,
By rebels and impostors daily vexed;
So many princes still were murderéd
The royal blood was nigh extinguishéd.
Then Tigranes, the great Armenian king,
To take the government was calléd in;
Lucullus him — the Roman general —
Vanquished in fight, and took those kingdoms all.
Of Greece and Syria thus the rule did end.
In Egypt next a little time we 'll spend.
First, Ptolemy being dead, his famous son
Called Philadelphus did possess the throne,
At Alexandria a library did build,
And with seven hundred thousand volumes filled.
The seventy-two interpreters did seek
They might translate the Bible into Greek.
His son was Evergetes, the last prince
That valor showed, virtue, or excellence.
Philopator was Evergetes' son.
After, Epiphanes sat on the throne,
Philometor, Evergetes again,
And after him did false Lathyrus reign;
Then Alexander in Lathyrus' stead;
Next, Auletes, who cut off Pompey's head.
To all these names we Ptolemy must add,
For since the first they still that title had.

Fair Cleopatra next, last of that race,
Whom Julius Cæsar set in royal place,
She, with her paramour, Marc Antony,
Held for a time the Egyptian monarchy,
Till great Augustus had with him a fight
At Actium, where his navy was put to flight;
He, seeing his honor lost, his kingdom end,
Did by his sword his life soon after send.
His brave virago asps sets to her arms
To take her life and quit her from all harms;
For 't was not death nor danger she did dread,
But some disgrace in triumph to be led.
Here ends at last the Grecian monarchy,
Which by the Romans had its destiny.
Thus kings and kingdoms have their times and dates,
Their standings, overturnings, bounds, and fates;
Now up, now down, now chief, and then brought under.
The heavens thus rule to fill the world with wonder.
The Assyrian monarchy long time did stand,
But yet the Persian got the upper hand;
The Grecian them did utterly subdue,
And millions were subjected unto few.
The Grecian longer than the Persian stood,
Then came the Roman like a raging flood,
And with the torrent of his rapid course
Their crowns, their titles, riches, bears by force.

The first was likened to a head of gold;
Next, arms and breast of silver to behold;
The third, belly and thighs of brass in sight;
And last was iron, which breaketh all with might.
The stone out of the mountain then did rise
And smote those feet, those legs, those arms, and
  thighs,
Then gold, silver, brass, iron, and all the store
Became like chaff upon the threshing floor.
The first a lion, second was a bear,
The third a leopard which four wings did rear;
The last more strong and dreadful than the rest,
Whose iron teeth devoured every beast,
And when he had no appetite to eat
The residue he stamped under his feet.
Yet shall this lion, bear, this leopard, ram,
All trembling stand before the powerful Lamb.
With these three monarchies now have I done.
But how the fourth their kingdoms from them won,
And how from small beginnings it did grow
To fill the world with terror and with woe,
My tired brain leaves to some better pen.
This task befits not women like to men.
For what is past I blush excuse to make,
But humbly stand some grave reproof to take.
Pardon to crave for errors is but vain;
The subject was too high beyond my strain.

To frame apology for some offense
Converts our boldness into impudence.
This my presumption some now to requite,
Ne sutor ultra crepidam may write.

The End of the Grecian Monarchy.

After some days of rest my restless heart
To finish what 's begun new thoughts impart,
And maugre all resolves my fancy wrought
This fourth to the other three now might be brought.
Shortness of time and inability
Will force me to a confused brevity;
Yet in this chaos one shall easily spy
The vast limbs of a mighty monarchy.
Whate' er is found amiss take in good part
As faults proceeding from my head, not heart.

THE ROMAN MONARCHY, BEING THE FOURTH AND LAST, BEGINNING ANNO MUNDI 3213.

Stout Romulus, Rome's founder and first king,
Whom vestal Rhea to the world did bring,
His father was not Mars, as some devised,
But Amulius in armor all disguised.
Thus he deceived his niece she might not know
The double injury he then did do.
Where shepherds once had coats and sheep their folds,
Where swains and rustic peasants kept their holds,
A city fair did Romulus erect,
The mistress of the world in each respect.
His brother Remus there by him was slain
For leaping o'er the wall with some disdain.
The stones at first were cemented with blood,
And bloody hath it proved since first it stood.
This city built, and sacrifices done,
A form of government he next begun;
A hundred senators he likewise chose,
And with the style of Patres honored those.
His city to replenish men he wants;
Great privileges then to all he grants
That will within those strong built walls reside,
And this new gentle government abide.
Of wives there was so great a scarcity
They to their neighbors sue for a supply.

But all disdain alliance then to make,
So Romulus was forced this course to take:
Great shows he makes at tilt and tournament;
To see these sports the Sabines all are bent;
Their daughters by the Romans then were caught.
Then to recover them a field was fought;
But in the end to final peace they come,
And Sabines as one people dwelt in Rome.
The Romans now more potent begin to grow,
And Fidenates they wholly overthrow.
But Romulus then comes unto his end,
Some feigning to the gods he did ascend;
Others the seven and thirtieth of his reign
Affirm that by the Senate he was slain.

NUMA POMPILIUS.

Numa Pompilius next chose they king,
Held for his piety some sacred thing.
To Janus he that famous temple built
Kept shut in peace, set ope when blood was spilt;
Religious rites and customs instituted,
And priests and flamens likewise he deputed,
Their augurs strange, their gestures and attire,
And vestal maids to keep the holy fire.
The nymph Ægeria this to him told,
So to delude the people he was bold.
Forty-three years he ruled with generous praise,
Accounted for a god in after days.

TULLUS HOSTILIUS.

Tullus Hostilius was third Roman king,
Who martial discipline in use did bring.
War with the ancient Albans he did wage.
This strife to end six brothers did engage,
Three called Horatii on the Romans' side,
And Curiatii three Albans provide.
The Romans conquer, the others yield the day,
Yet in their compact after false they play.
The Romans, sore incensed, their general slay,
And from old Alba fetch the wealth away.
Of Latin kings this was long since the seat,
But now demolishéd to make Rome great.
Thirty-two years did Tullus reign, then die,
Left Rome in wealth and power still growing high.

ANCUS MARCIUS.

Next Ancus Marcius sits upon the throne,
Nephew unto Pompilius dead and gone.
Rome he enlarged, new built again the wall
Much stronger, and more beautiful withal.
A stately bridge he over Tiber made,
Of boats and oars no more they need the aid.
Fair Ostia he built; this town it stood
Close by the mouth of famous Tiber flood.
Twenty-four years time of his royal race,
Then unto death unwillingly gives place.

TARQUINIUS PRISCUS.

Tarquin, a Greek at Corinth born and bred,
Who from his country for sedition fled,
Is entertained at Rome, and in short time
By wealth and favor doth to honor climb.
He after Marcius' death the kingdom had.
A hundred senators he more did add.
Wars with the Latins he again renews,
And nations twelve of Tuscany subdues.
To such rude triumphs as young Rome then had
Some state and splendor did this Priscus add.
Thirty-eight years this stranger born did reign,
And after all by Ancus' sons was slain.

SERVIUS TULLIUS.

Next Servius Tullius gets into the throne;
Ascends not up by merits of his own,
But by the favor and the special grace
Of Tanaquil, late queen, obtains the place.
He ranks the people into each degree
As wealth had made them of ability;
A general muster takes, which by account
To eighty thousand souls then did amount.
Forty-four years did Servius Tullius reign,
And then by Tarquin Priscus' son was slain.

TARQUINIUS SUPERBUS, THE LAST KING OF THE ROMANS.

Tarquin the proud, from manners calléd so,
Sat on the throne when he had slain his foe.
Sextus, his son, did most unworthily
Lucretia force, mirror of chastity;
She loathéd so the fact, she loathed her life,
And shed her guiltless blood with guilty knife.
Her husband, sore incensed to quit this wrong,
With Junius Brutus rose, and being strong
The Tarquins they from Rome by force expel,
In banishment perpetual to dwell;
The government they change, a new one bring,
And people swear ne'er to accept of king.

An Apology.

To finish what's begun was my intent,
My thoughts and my endeavors thereto bent;
Essays I many made, but still gave out,
The more I mused, the more I was in doubt —
The subject large, my mind and body weak,
With many more discouragements did speak.
All thoughts of further progress laid aside,
Though oft persuaded, I as oft denied.
At length resolved, when many years had passed,
To prosecute my story to the last;
And for the same I hours not few did spend,
And weary lines, though lank, I many penned.
But 'fore I could accomplish my desire
My papers fell a prey to the raging fire,
And thus my pains, with better things, I lost,
Which none had cause to wail, nor I to boast.
No more I'll do, since I have suffered wreck,
Although my Monarchies their legs do lack,
Nor matter is it this last, the world now sees,
Hath many ages been upon his knees.

A DIALOGUE BETWEEN OLD ENGLAND AND NEW CONCERNING THEIR PRESENT TROUBLES, ANNO 1642.

NEW-ENGLAND.

Alas, dear mother, fairest queen and best,
With honor, wealth, and peace happy and blest,
What ails thee hang thy head, and cross thine arms,
And sit in the dust to sigh these sad alarms?
What deluge of new woes thus overwhelms
The glories of thy ever famous realm?
What means this wailing tone, this mournful guise?
Ah, tell thy daughter, she may sympathize.

OLD ENGLAND.

Art ignorant indeed of these my woes,
Or must my forcéd tongue these griefs disclose,
And must myself dissect my tattered state,
Which amazéd Christendom stands wondering at?
And thou a child, a limb, and dost not feel
My fainting, weakened body now to reel?

REV. JOHN COTTON.
The most influential of the early colonial divines.
The Boston of the New World was named
in memory of his English home.
From an engraving.

This physic purging potion I have taken
Will bring consumption or an ague-quaking
Unless some cordial thou fetch from high,
Which present help may ease my malady.
If I decease, dost think thou shalt survive?
Or by my wasting state dost think to thrive?
Then weigh our case if it be not justly sad.
Let me lament alone, while thou art glad.

NEW-ENGLAND.

And thus, alas, your state you much deplore
In general terms, but will not say wherefore.
What medicine shall I seek to cure this woe
If the wound so dangerous I may not know.
But you, perhaps, would have me guess it out.
What, hath some Hengist like that Saxon stout
By fraud or force usurped thy flowering crown,
Or by tempestuous wars thy fields trod down?
Or hath Canutus, that brave valiant Dane,
The regal peaceful scepter from thee ta'en?
Or is it a Norman whose victorious hand
With English blood bedews thy conquered land?
Or is it intestine wars that thus offend?
Do Maud and Stephen for the crown contend?
Do barons rise and side against their king,
And call in foreign aid to help the thing?
Must Edward be deposed? Or is it the hour
That second Richard must be clapped in the tower?

Or is it the fatal jar, again begun,
That from the red-white pricking roses sprung?
Must Richmond's aid the nobles now implore
To come and break the tushes of the boar?
If none of these, dear mother, what's your woe?
Pray, do you fear Spain's bragging Armado?
Doth your ally, fair France, conspire your wreck,
Or do the Scots play false behind your back?
Doth Holland quit you ill for all your love?
Whence is the storm, from earth or heaven above?
Is it drought, is it famine, or is it pestilence?
Dost feel the smart, or fear the consequence?
Your humble child entreats you show your grief.
Though arms nor purse she hath for your relief,—
Such is her poverty,— yet shall be found
A suppliant for your help, as she is bound.

OLD ENGLAND.

I must confess some of those sores you name
My beauteous body at this present maim;
But foreign foe nor feignéd friend I fear,
For they have work enough, thou knowest, elsewhere.
Nor is it Alcie's son nor Henry's daughter
Whose proud contentions cause this slaughter;
Nor nobles siding to make John no king,
French Louis unjustly to the crown to bring;
No Edward, Richard, to lose rule and life,
Nor no Lancastrians to renew old strife;

A Dialogue between Old England and New

No Duke of York nor Earl of March to soil
Their hands in kindred's blood whom they did foil.
No crafty tyrant now usurps the seat
Who nephews slew that so he might be great.
No need of Tudor roses to unite;
None knows which is the red or which the white.
Spain's braving fleet a second time is sunk.
France knows how oft my fury she hath drunk
By Edward Third and Henry Fifth of fame;
Her lilies in my arms avouch the same.
My sister Scotland hurts me now no more,
Though she hath been injurious heretofore.
What Holland is I am in some suspense,
But trust not much unto his excellence.
For wants, sure some I feel, but more I fear;
And for the pestilence, who knows how near?
Famine and plague, two sisters of the sword,
Destruction to a land doth soon afford.
They 're for my punishment ordained on high,
Unless our tears prevent it speedily.
But yet I answer not what you demand
To show the grievance of my troubled land.
Before I tell the effect I 'll show the cause,
Which is my sins — the breach of sacred laws:
Idolatry, supplanter of a nation,
With foolish superstitious adoration,
Are liked and countenanced by men of might;
The gospel trodden down and hath no right;

Church offices were sold and bought for gain,
That Pope had hope to find Rome here again;
For oaths and blasphemies did ever ear
From Beelzebub himself such language hear?
What scorning of the saints of the most high!
What injuries did daily on them lie!
What false reports, what nicknames did they take,
Not for their own, but for their Master's sake!
And thou, poor soul, wert jeered among the rest;
Thy flying for the truth was made a jest.
For Sabbath-breaking and for drunkenness
Did ever land profaneness more express?
From crying blood yet cleansèd am not I,
Martyrs and others dying causelessly.
How many princely heads on blocks laid down
For naught but title to a fading crown!
'Mongst all the cruelties by great ones done,
O Edward's youths, and Clarence' hapless son,
O Jane, why didst thou die in flowering prime?—
Because of royal stem, that was thy crime.
For bribery, adultery, and lies
Where is the nation I can't paralyze?
With usury, extortion, and oppression,
These be the hydras of my stout transgression;
These be the bitter fountains, heads, and roots
Whence flowed the source, the sprigs, the boughs, and fruits.

A Dialogue between Old England and New

Of more than thou canst hear or I relate,
That with high hand I still did perpetrate.
For these were threatenéd the woeful day.
I mocked the preachers, put it far away;
The sermons yet upon record do stand
That cried destruction to my wicked land.
I then believed not, now I feel and see
The plague of stubborn incredulity.
Some lost their livings, some in prison pent,
Some, fined, from house and friends to exile went.
Their silent tongues to heaven did vengeance cry,
Who saw their wrongs, and hath judged righteously,
And will repay it sevenfold in my lap.
This is forerunner of my afterclap.
Nor took I warning by my neighbors' falls:
I saw sad Germany's dismantled walls,
I saw her people famished, nobles slain,
Her fruitful land a barren heath remain;
I saw, unmoved, her armies foiled and fled,
Wives forced, babes tossed, her houses calcinéd.
I saw strong Rochelle yielded to her foe,
Thousands of starvéd Christians there also.
I saw poor Ireland bleeding out her last,
Such cruelties as all reports have passed;
Mine heart obdurate stood not yet aghast.
Now sip I of that cup, and just it may be
The bottom dregs reservéd are for me.

NEW-ENGLAND.

To all you 've said, sad mother, I assent.
Your fearful sins great cause there is to lament.
My guilty hands in part hold up with you,
A sharer in your punishment 's my due.
But all you say amounts to this effect,
Not what you feel, but what you do expect.
Pray, in plain terms, what is your present grief?
Then let 's join heads and hearts for your relief.

OLD ENGLAND.

Well, to the matter, then. There 's grown of late
'Twixt king and peers a questiön of state:
Which is the chief— the law, or else the king?
One said, it 's he; the other, no such thing.
'T is said my better part in Parliament
To ease my groaning land showed their intent,
To crush the proud, and right to each man deal,
To help the church, and stay the commonweal.
So many obstacles came in their way
As puts me to a stand what I should say.
Old customs new prerogatives stood on;
Had they not held law fast, all had been gone,
Which by their prudence stood them in such stead
They took high Strafford lower by the head,
And to their *Laud* be it spoke they held in the tower
All England's metropolitan that hour.

JOHN WINTHROP.
The First Governor of the Massachusetts Bay Colony.
From the original painting in the State House, Boston, Mass.

A Dialogue between Old England and New

This done, an act they would have passéd fain
No prelate should his bishopric retain;
Here tugged they hard indeed, for all men saw
This must be done by gospel, not by law.
Next the militiā they urgéd sore;
This was denied, I need not say wherefore.
The king, displeased, at York himself absents.
They humbly beg his return, show their intents;
The writing, printing, posting to and fro,
Show all was done; I'll therefore let it go.
But now I come to speak of my disaster.
Contention grown 'twixt subjects and their master,
They worded it so long they fell to blows,
That thousands lay on heaps. Here bleed my woes.
I that no wars so many years have known
Am now destroyed and slaughtered by my own.
But could the field alone this strife decide,
One battle, two, or three I might abide.
But these may be beginnings of more woe —
Who knows but this may be my overthrow!
Oh, pity me in this sad perturbation,
My plundered towns, my houses' devastation,
My weeping virgins, and my young men slain,
My wealthy trading fallen, my dearth of grain.
The seed-times come, but plowman hath no hope
Because he knows not who shall in his crop.
The poor they want their pay, their children bread,
Their woeful mothers' tears unpitiéd.

If any pity in thy heart remain,
Or any child-like love thou dost retain,
For my relief do what there lies in thee,
And recompense that good I've done to thee.

NEW ENGLAND.

Dear mother, cease complaints, and wipe your eyes,
Shake off your dust, cheer up, and now arise.
You are my mother nurse, and I, your flesh,
Your sunken bowels gladly would refresh.
Your griefs I pity, but soon hope to see
Out of your troubles much good fruit to be;
To see those latter days of hoped-for good,
Though now beclouded all with tears and blood.
After dark popery the day did clear;
But now the sun in his brightness shall appear.
Blest be the nobles of thy noble land
With ventured lives for truth's defense that stand.
Blest be thy Commons, who for common good
And thy infringéd laws have boldly stood.
Blest be thy counties, who did aid thee still
With hearts and states to testify their will.
Blest be thy preachers, who do cheer thee on;
Oh, cry the sword of God and Gideon!
And shall I not on them wish Meroz' curse
That help thee not with prayers, with alms, and purse?
And for myself let miseries abound
If mindless of thy state I e'er be found.

These are the days the church's foes to crush,
To root out popeling's head, tail, branch, and rush.
Let's bring Baal's vestments forth to make a fire,
Their miters, surplices, and all their attire,
Copes, rochets, croziers, and such empty trash,
And let their names consume, but let the flash
Light Christendom, and all the world to see
We hate Rome's whore, with all her trumpery.
Go on, brave Essex, with a loyal heart,
Not false to king, nor to the better part;
But those that hurt his people and his crown,
As duty binds expel and tread them down.
And ye brave nobles, chase away all fear,
And to this hopeful cause closely adhere.
O mother, can you weep and have such peers?
When they are gone, then drown yourself in tears,
If now you weep so much, that then no more
The briny ocean will o'erflow your shore.
These, these are they, I trust, with Charles our king,
Out of all mists such glorious days shall bring
That dazzled eyes, beholding, much shall wonder
At that thy settled peace, thy wealth, and splendor;
Thy church and weal established in such manner
That all shall joy that thou displayedst thy banner;
And discipline erected so, I trust,
That nursing kings shall come and lick thy dust.
Then justice shall in all thy courts take place
Without respect of person or of case;

Then bribes shall cease, and suits shall not stick long,
Patience and purse of clients oft to wrong;
Then high commissions shall fall to decay,
And pursuivants and catchpoles want their pay.
So shall thy happy nation ever flourish,
When truth and righteousness they thus shall nourish.
When thus in peace, thine armies brave send out
To sack proud Rome, and all her vassals rout;
There let thy name, thy fame, and glory shine,
As did thine ancestors' in Palestine,
And let her spoils full pay with interest be
Of what unjustly once she polled from thee.
Of all the woes thou canst let her be sped,
And on her pour the vengeance threatenéd.
Bring forth the beast that ruled the world with his beck,
And tear his flesh, and set your feet on his neck,
And make his filthy den so desolate
To the astonishment of all that knew his state.
This done, with brandished swords to Turkey go,—
For then what is it but English blades dare do?—
And lay her waste,— for so 's the sacred doom,—
And do to Gog as thou hast done to Rome.
O Abraham's seed, lift up your heads on high,
For sure the day of your redemption 's nigh.
The scales shall fall from your long blinded eyes,
And him you shall adore who now despise.
Then fulness of the nations in shall flow,
And Jew and Gentile to one worship go;

JOHN ELIOT.
"The Apostle of the Indians."
From the original painting owned by Mrs. William Whiting,
Roxbury, Mass

A Dialogue between Old England and New

Then follow days of happiness and rest.
Whose lot doth fall to live therein is blest.
No Canaanite shall then be found in the land,
And holiness on horses' bells shall stand.
If this make way thereto, then sigh no more,
But if at all thou didst not see it before.
Farewell, dear mother; rightest cause prevail,
And in a while you 'll tell another tale.

AN ELEGY UPON THAT HONORABLE AND RENOWNED KNIGHT SIR PHILIP SIDNEY, WHO WAS UNTIMELY SLAIN AT THE SIEGE OF ZÜTPHEN, ANNO 1586.

When England did enjoy her halcyon days
Her noble Sidney wore the crown of bays,
As well an honor to our British land
As she that swayed the scepter with her hand.
Mars and Minerva did in one agree
Of arms and arts he should a pattern be;
Calliope with Terpsichore did sing
Of poesy and of music he was king.
His rhetoric struck Polyhymnia dead,
His eloquence made Mercury wax red,
His logic from Euterpe won the crown,
More worth was his than Clio could set down.
Thalia and Melpomene, say the truth,—
Witness "Arcadia" penned in his youth,—
Are not his tragic comedies so acted
As if your ninefold wit had been compacted

SIR PHILIP SIDNEY.

An Elegy upon Sir Philip Sidney

To show the world they never saw before,
That this one volume should exhaust your store?
His wiser days condemned his witty works.
Who knows the spell that in his rhetoric lurks?
But some infatuate fools, soon caught therein,
Found Cupid's dam had never such a gin,
Which makes severer eyes but slight that story,
And men of morose minds envy his glory.
But he 's a beetlehead that can't descry
A world of wealth within that rubbish lie,
And doth his name, his work, his honor wrong,
The brave refiner of our British tongue,
That sees not learning, valor, and morality,
Justice, friendship, and kind hospitality,
Yea, and divinity, within his book.
Such were prejudicate, and did not look.
In all records his name I ever see
Put with an epithet of dignity,
Which shows his worth was great, his honor such
The love his country ought him was as much.
Then let none disallow of these my strains
Whilst English blood yet runs within my veins.
O brave Achilles, I wish some Homer would
Engrave in marble, with characters of gold,
The valiant feats thou didst on Flanders' coast,
Which at this day fair Belgia may boast.
The more I say, the more thy worth I stain.
Thy fame and praise are far beyond my strain.

O Zütphen, Zütphen, that most fatal city
Made famous by thy death, much more the pity!
Ah! in his blooming prime death plucked this rose
Ere he was ripe, his thread cut Atropos.
Thus man is born to die, and dead is he.
Brave Hector by the walls of Troy we see.
Oh, who was near thee but did sore repine
He rescued not with life that life of thine?
But yet impartial fates this boon did give —
Though Sidney died, his valiant name should live.
And live it doth, in spite of death through fame.
Thus being overcome, he overcame.
Where is that envious tongue but can afford
Of this our noble Scipio some good word?
Great Bartas, this unto thy praise adds more,
In sad sweet verse thou didst his death deplore.
And phenix Spenser doth unto his life
His death present in sable to his wife,
Stella the fair, whose streams from conduits fell
For the sad loss of her dear Astrophel.
Fain would I show how he fame's paths did tread,
But now into such labyrinths I am led,
With endless turns, the way I find not out.
How to persist, my muse is more in doubt,
Which makes me now with Sylvester confess
But Sidney's muse can sing his worthiness.
The Muses' aid I craved; they had no will
To give to their detractor any quill.

An Elegy upon Sir Philip Sidney

With high disdain they said they gave no more
Since Sidney had exhausted all their store.
They took from me the scribbling pen I had;
I to be eased of such a task was glad;
Then to revenge this wrong themselves engage,
And drave me from Parnassus in a rage.
Then wonder not if I no better sped,
Since I the Muses thus have injuréd.
I, pensive for my fault, sat down, and then
Errata, through their leave, threw me my pen;
My poem to conclude two lines they deign,
Which writ, she bade return it to them again.
So Sidney's fame I leave to England's rolls.
His bones do lie interred in stately Paul's.

His Epitaph.

Here lies in fame under this stone
Philip and Alexander both in one,
Heir to the Muses, the son of Mars in truth,
Learning, valor, wisdom, all in virtuous youth.
His praise is much; this shall suffice my pen
That Sidney died 'mong most renowned of men.

IN HONOR OF DU BARTAS, 1641.

Among the happy wits this age hath shown,
Great, dear, sweet Bartas, thou art matchless known.
My ravished eyes and heart, with faltering tongue,
In humble wise have vowed their service long,
But knowing the task so great, and strength but small,
Gave o'er the work before begun withal.
My dazzled sight of late reviewed thy lines,
Where art, and more than art, in nature shines.
Reflection from their beaming altitude
Did thaw my frozen heart's ingratitude,
Which rays, darting upon some richer ground,
Had caused flowers and fruits soon to abound;
But barren I my daisy here do bring,
A homely flower in this my latter spring.
If summer or my autumn age do yield
Flowers, fruits, in garden, orchard, or in field,
They shall be consecrated in my verse
And prostrate offered at great Bartas' hearse.
My muse unto a child I may compare
Who sees the riches of some famous fair;

WILLIAM SALLUST DU BARTAS.
From an engraving.

In Honor of Du Bartas

He feeds his eyes, but understanding lacks
To comprehend the worth of all those knacks.
The glittering plate and jewels he admires,
The hats and fans, the plumes and ladies' attires,
And thousand times his amazéd mind doth wish
Some part, at least, of that brave wealth were his;
But seeing empty wishes naught obtain,
At night turns to his mother's cot again,
And tells her tales, his full heart over-glad,
Of all the glorious sights his eyes have had,
But finds too soon his want of eloquence.
The silly prattler speaks no word of sense,
But seeing utterance fail his great desires,
Sits down in silence, deeply he admires.
Thus weak-brained I, reading thy lofty style,
Thy profound learning, viewing other while
Thy art in natural philosophy,
Thy saint-like mind in grave divinity,
Thy piercing skill in high astronomy,
And curious insight in anatomy,
Thy physic, music, and state policy,
Valor in war, in peace good husbandry.
Sure liberal nature did with art not small
In all the arts make thee most liberal.
A thousand thousand times my senseless senses
Moveless stand, charmed by thy sweet influences,
More senseless than the stones to Amphion's lute;
Mine eyes are sightless, and my tongue is mute,

My full astonished heart doth pant to break,
Through grief it wants a faculty to speak.
Volleys of praises could I echo then
Had I an angel's voice or Bartas' pen.
But wishes can't accomplish my desire.
Pardon if I adore when I admire.
O France, thou didst in him more glory gain
Than in thy Martel, Pépin, Charlemagne,
Than in St. Louis, or thy last Henry great,
Who tamed his foes in wars, in blood, and sweat.
Thy fame is spread as far, I dare be bold,
In all the zones, the temperate, hot, and cold.
Their trophies were but heaps of wounded slain;
Thine the quintessence of an heroic brain.
The oaken garland ought to deck their brows;
Immortal bays to thee all men allow,
Who in thy triumphs never won by wrongs,
Lead'st millions chained by eyes, by ears, by tongues.
Oft have I wondered at the hand of heaven
In giving one what would have servéd seven.
If e'er this golden gift were showered on any,
Thy double portion would have servéd many.
Unto each man his riches are assigned
Of name, of state, of body, and of mind;
Thou hadst thy part of all but of the last.
O pregnant brain, O comprehension vast,
Thy haughty style and rapted wit sublime
All ages, wondering at, shall never climb.

Thy sacred works are not for imitation,
But monuments to future admiration.
Thus Bartas' fame shall last while stars do stand,
And whilst there 's air, or fire, or sea, or land.
But lest mine ignorance should do thee wrong,
To celebrate thy merits in my song,
I 'll leave thy praise to those shall do thee right.
Good-will, not skill, did cause me bring my mite.

His Epitaph.

Here lies the Pearl of France, Parnassus' glory;
The world rejoiced at his birth, at his death was sorry.
Art and Nature joined by heaven's high decree
Now showed what once they ought, humanity;
And Nature's law had it been revocable
To rescue him from death Art had been able.
But Nature vanquished Art, so Bartas died;
But fame outliving both, he is revived.

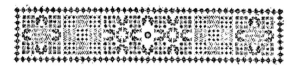

IN HONOR OF THAT HIGH AND MIGHTY PRINCESS QUEEN ELIZABETH OF HAPPY MEMORY.

THE PROEM.

Although, great queen, thou now in silence lie,
Yet thy loud herald, fame, doth to the sky
Thy wondrous worth proclaim in every clime,
And so hath vowed while there is world or time.
So great is thy glory and thine excellence
The sound thereof rapts every human sense,
That men account it no impiety
To say thou wert a fleshly deity.
Thousands bring offerings, though out of date,
Thy world of honors to accumulate;
'Mongst hundred hecatombs of roaring verse,
Mine bleating stands before thy royal hearse.
Thou never didst nor canst thou now disdain
To accept the tribute of a loyal brain;
Thy clemency did erst esteem as much
The acclamations of the poor as rich,
Which makes me deem my rudeness is no wrong,
Though I resound thy praises 'mongst the throng.

THE POEM.

No phenix pen, nor Spenser's poetry,
Nor Speed's nor Camden's learned history,
Eliza's works, wars, praise, can e'er compact.
The world 's the theater where she did act.
No memories nor volumes can contain
The eleven olympiads of her happy reign,
Who was so good, so just, so learned, wise,
From all the kings on earth she won the prize.
Nor say I more than duly is her due;
Millions will testify that this is true.
She hath wiped off the aspersion of her sex
That women wisdom lack to play the rex.
Spain's monarch says not so, nor yet his host;
She taught them better manners to their cost.
The salic law in force now had not been
If France had ever hoped for such a queen.
But can you, doctors, now this point dispute,
She 's argument enough to make you mute.
Since first the sun did run his near-run race,
And earth had, once a year, a new-old face,
Since time was time, and man unmanly man,
Come show me such a phenix if you can?
Was ever people better ruled than hers?
Was ever land more happy, freed from stirs?
Did ever wealth in England more abound?
Her victories in foreign coasts resound.

Ships more invincible than Spain's, her foe
She wrecked, she sacked, she sunk his Armado;
Her stately troops advanced to Lisbon's wall
Don Anthony in his right there to install;
She frankly helped Franks' brave distresséd king;
The states united now her fame do sing,
She their protectrix was — they well do know
Unto our dread virago what they owe.
Her nobles sacrificed their noble blood,
Nor men nor coin she spared to do them good.
The rude untaméd Irish she did quell;
Before her picture the proud Tyrone fell.
Had ever prince such counsellors as she?
Herself, Minerva, caused them so to be.
Such captains and such soldiers never seen
As were the subjects of our Pallas queen.
Her seamen through all straits the world did round
Terra incognita might know the sound.
Her Drake came laden home with Spanish gold;
Her Essex took Cadiz, their herculean hold.
But time would fail me, so my tongue would, too,
To tell of half she did or she could do.
Semiramis to her is but obscure —
More infamy than fame she did procure;
She built her glory but on Babel's walls,
World's wonder for awhile, but yet it falls.
Fierce Tomyris (Cyrus' headsman), Scythians' queen,
Had put her harness off had she but seen

In Honor of Queen Elizabeth

Our amazon in the camp of Tilbury,
Judging all valor and all majesty
Within that princess to have residence,
And prostrate yielded to her excellence.
Dido, first foundress of proud Carthage' walls,—
Who living consummates her funerals?—
A great Elisa; but compared with ours
How vanisheth her glory, wealth, and powers!
Profuse, proud Cleopatra, whose wrong name,
Instead of glory, proved her country's shame,
Of her what worth in stories to be seen
But that she was a rich Egyptian queen?
Zenobia, potent empress of the East,
And of all these without compare the best,
Whom none but great Aurelian could quell,
Yet for our queen is no fit parallel.
She was a phenix queen; so shall she be,
Her ashes not revived, more phenix she.
Her personal perfections who would tell
Must dip his pen in the Heliconian well,
Which I may not; my pride doth but aspire
To read what others write, and so admire.
Now say, have women worth, or have they none?
Or had they some, but with our queen is it gone?
Nay, masculines, you have thus taxed us long,
But she, though dead, will vindicate our wrong.
Let such as say our sex is void of reason
Know 't is a slander now, but once was treason.

But happy England, which had such a queen!
Yea, happy, happy, had those days still been!
But happiness lies in a higher sphere;
Then wonder not Eliza moves not here.
Full fraught with honor, riches, and with days,
She set, she set, like Titan in his rays.
No more shall rise or set so glorious sun
Until the heavens' great revolutiön.
If then new things their old forms shall retain,
Eliza shall rule Albion once again.

Her Epitaph.

Here sleeps the queen; this is the royal bed
Of the damask rose sprung from the white and red,
Whose sweet perfume fills the all-filling air.
This rose is withered, once so lovely fair.
On neither tree did grow such rose before;—
The greater was our gain, our loss the more.

Another.

Here lies the pride of queens, pattern of kings.
So blaze it, Fame; here are feathers for thy wings.
Here lies the envied yet unparalleled prince,
Whose living virtues speak, though dead long since.
If many worlds, as that fantastic framed,
In every one be her great glory famed.

DAVID'S LAMENTATION FOR SAUL AND JONATHAN.

II Samuel 1. 19.

Alas, slain is the head of Israel,
Illustrious Saul, whose beauty did excel!
Upon thy places mountainous and high
How did the mighty fall, and, falling, die!
In Gath let not these things be spoken on,
Nor published in the streets of Askelon,
Lest daughters of the Philistines rejoice,
Lest the uncircumcised lift up their voice.
O Gilboa mounts, let never pearléd dew
Nor fruitful showers your barren tops bestrew,
Nor fields of offerings ever on you grow,
Nor any pleasant thing e'er may you show;
For there the mighty ones did soon decay,
The shield of Saul was vilely cast away;
There had his dignity so sore a foil
As if his head ne'er felt the sacred oil.
Sometimes from crimson blood of ghastly slain
The bow of Jonathan ne'er turned in vain;

Nor from the fat and spoils of mighty men
With bloodless sword did Saul turn back again.
Pleasant and lovely were they both in life,
And in their death was found no parting strife.
Swifter than swiftest eagles so were they,
Stronger than lions ramping for their prey.
O Israel's dames, o'erflow your beauteous eyes
For valiant Saul, who on Mount Gilboa lies,
Who clothéd you in cloth of richest dye,
And choice delights full of variety,
On your array put ornaments of gold,
Which made you yet more beauteous to behold.
Oh, how in battle did the mighty fall
In midst of strength, not succoréd at all!
O lovely Jonathan, how wast thou slain!
In places high full low thou didst remain.
Distressed for thee I am, dear Jonathan;
Thy love was wonderful, surpassing man,
Exceeding all the love that 's feminine,
So pleasant hast thou been, dear brother mine.
How are the mighty fallen into decay,
And warlike weapons perishéd away!

TO THE MEMORY OF MY DEAR AND EVER HONORED FATHER THOMAS DUDLEY, ESQ., WHO DECEASED JULY 31, 1653, AND OF HIS AGE 77.

By duty bound, and not by custom led
To celebrate the praises of the dead,
My mournful mind, sore pressed, in trembling verse
Presents my lamentations at his hearse
Who was my father, guide, instructor, too,
To whom I ought whatever I could do.
Nor is it relation near my hand shall tie;
For who more cause to boast his worth than I?
Who heard, or saw, observed, or knew him better,
Or who alive than I a greater debtor?
Let malice bite, and envy gnaw its fill,
He was my father, and I'll praise him still.
Nor was his name or life led so obscure
That pity might some trumpeters procure,
Who after death might make him falsely seem
Such as in life no man could justly deem.
Well known and loved, where'er he lived, by most,
Both in his native and in foreign coast,
These to the world his merits could make known,
So need no testimonial from his own.

But now or never I must pay my sum;
While others tell his worth, I'll not be dumb.
One of thy founders him, New England, know,
Who stayed thy feeble sides when thou wast low,
Who spent his state, his strength, and years with care
That aftercomers in them might have share.
True patriot of this little commonweal,
Who is it can tax thee aught but for thy zeal?
Truth's friend thou wert, to errors still a foe,
Which caused apostates to malign thee so.
Thy love to true religion e'er shall shine —
My father's God be God of me and mine!
Upon the earth he did not build his nest,
But as a pilgrim what he had possessed.
High thoughts he gave no harbor in his heart,
Nor honors puffed him up, when he had part;
Those titles loathed which some too much do love,
For truly his ambition lay above.
His humble mind so loved humility
He left it to his race for legacy,
And oft and oft, with speeches mild and wise,
Gave his in charge that jewel rich to prize.
No ostentation seen in all his ways,
As in the mean ones of our foolish days,
Which all they have, and more, still set to view
Their greatness may be judged by what they show.
His thoughts were more sublime, his actions wise;
Such vanities he justly did despise.

Nor wonder 't was low things ne'er much did move,
For he a mansion had prepared above,
For which he sighed and prayed and longed full sore
He might be clothed upon for evermore;
Oft spake of death, and with a smiling cheer
He did exult his end was drawing near.
Now fully ripe, as shock of wheat that 's grown,
Death as a sickle hath him timely mown,
And in celestial barn hath housed him high,
Where storms, nor showers, nor aught can damnify.
His generation served, his labors cease,
And to his fathers gathered is in peace.
Ah happy soul, 'mongst saints and angels blest,
Who after all his toil is now at rest!
His hoary head in righteousness was found;
As joy in heaven, on earth let praise resound.
Forgotten never be his memory!
His blessing rest on his posterity!
His pious footsteps followed by his race
At last will bring us to that happy place
Where we with joy each other's face shall see,
And parted more by death shall never be.

His Epitaph.

Within this tomb a patriot lies
That was both pious, just, and wise,
To truth a shield, to right a wall,
To sectaries a whip and maul.

A magazine of history,
A prizer of good company,
In manners pleasant and severe,
The good him loved, the bad did fear;
And when his time with years was spent,
If some rejoiced, more did lament.

AN EPITAPH ON MY DEAR AND EVER HONORED MOTHER MRS. DOROTHY DUDLEY, WHO DECEASED DECEMBER 27, 1643, AND OF HER AGE 61.

Here lies
A worthy matron of unspotted life,
A loving mother, and obedient wife,
A friendly neighbor, pitiful to poor,
Whom oft she fed and clothéd with her store;
To servants wisely aweful, but yet kind,
And as they did so they reward did find;
A true instructor of her family,
The which she ordered with dexterity;
The public meetings ever did frequent,
And in her closet constant hours she spent;
Religiöus in all her words and ways,
Preparing still for death till end of days;
Of all her children children lived to see,
Then, dying, left a blessed memory.

CONTEMPLATIONS.

Some time now past in the autumnal tide,
    When Phœbus wanted but one hour to bed,
The trees all richly clad, yet void of pride,
    Were gilded o'er by his rich golden head;
Their leaves and fruits seemed painted, but were true
Of green, of red, of yellow, mixéd hue.
Rapt were my senses at this delectable view.

I wist not what to wish, yet sure, thought I,
    If so much excellence abide below
How excellent is He that dwells on high,
    Whose power and beauty by his works we know!
Sure He is goodness, wisdom, glory, light,
That hath this under world so richly dight.
More heaven than earth was here, no winter and no night.

Then on a stately oak I cast mine eye,
    Whose ruffling top the clouds seemed to aspire.
How long since thou wast in thine infancy?
    Thy strength and stature more thy years admire.

Hath hundred winters passed since thou wast born,
Or thousand since thou break'st thy shell of horn?
If so, all these as naught eternity doth scorn.

Then higher on the glistering sun I gazed,
　　Whose beams were shaded by the leafy tree;
The more I looked the more I grew amazed,
　　And softly said, What glory 's like to thee?
Soul of this world, this universe's eye,
No wonder some made thee a deity.
Had I not better known, alas, the same had I.

Thou as a bridegroom from thy chamber rushes,
　　And as a strong man joys to run a race;
The morn doth usher thee with smiles and blushes,
　　The earth reflects her glances in thy face.
Birds, insects, animals, with vegetive,
Thy heart from death and dulness doth revive,
And in the darksome womb of fruitful nature dive.

Thy swift annual and diurnal course,
　　Thy daily straight and yearly oblique path,
Thy pleasing fervor, and thy scorching force
　　All mortals here the feeling knowledge hath.
Thy presence makes it day, thy absence night.
Quaternal seasons caused by thy might.
Hail, creature full of sweetness, beauty, and delight!

Art thou so full of glory that no eye
    Hath strength thy shining rays once to behold?
And is thy splendid throne erect so high
    As to approach it can no earthly mould?
How full of glory then must thy Creator be
Who gave this bright light luster unto thee?
Admired, adored, forever be that Majesty.

Silent, alone, where none or saw or heard,
    In pathless paths I led my wandering feet;
My humble eyes to lofty skies I reared,
    To sing some song my amazéd muse thought meet.
My great Creator I would magnify
That nature had thus decked liberally.
But ah, and ah again, my imbecility!

I heard the merry grasshopper then sing,
    The black-clad cricket bear a second part;
They kept one tune and played on the same string,
    Seeming to glory in their little art.
Shall creatures abject thus their voices raise,
And in their kind resound their maker's praise,
Whilst I as mute can warble forth no higher lays?

When present times look back to ages past,
    And men in being fancy those are dead,
It makes things gone perpetually to last,
    And calls back months and years that long since fled;
It makes a man more aged in conceit
Than was Methuselah or his grandsire great
While of their persons and their acts his mind doth treat.

Sometimes in Eden fair he seems to be,
    Sees glorious Adam there made lord of all,
Fancies the apple dangle on the tree
    That turned his sovereign to a naked thrall,
Who like a miscreant was driven from that place
To get his bread with pain and sweat of face —
A penalty imposed on his backsliding race.

Here sits our grandam in retired place,
    And in her lap her bloody Cain new born;
The weeping imp oft looks her in the face,
    Bewails his unknown hap and fate forlorn.
His mother sighs to think of paradise,
And how she lost her bliss to be more wise,
Believing him that was and is father of lies.

Here Cain and Abel come to sacrifice;
    Fruits of the earth and fatlings each doth bring.
On Abel's gift the fire descends from skies,
    But no such sign on false Cain's offering.

With sullen hateful looks he goes his ways,
Hath thousand thoughts to end his brother's days,
Upon whose blood his future good he hopes to raise.

There Abel keeps his sheep, no ill he thinks;
   His brother comes, then acts his fratricide;
The virgin earth of blood her first draught drinks,
   But since that time she often hath been cloyed.
The wretch with ghastly face and dreadful mind
Thinks each he sees will serve him in his kind,
Though none on earth but kindred near then could
     he find.

Who fancies not his looks now at the bar,
   His face like death, his heart with horror fraught,
Nor malefactor ever felt like war
   When deep despair with wish of life hath sought.
Branded with guilt, and crushed with treble woes,
A vagabond to land of Nod he goes,
A city builds, that walls might him secure from foes.

Who thinks not oft upon the fathers' ages,
   Their long descent, how nephews' sons they saw,
The starry observations of those sages,
   And how their precepts to their sons were law;
How Adam sighed to see his progeny
Clothed all in his black sinful livery,
Who neither guilt nor yet the punishment could fly?

Our life compare we with their length of days,
  Who to the tenth of theirs doth now arrive?
And though thus short, we shorten many ways,
  Living so little while we are alive —
In eating, drinking, sleeping, vain delight;
So unawares comes on perpetual night,
And puts all pleasures vain unto eternal flight.

When I behold the heavens as in their prime,
  And then the earth, though old, still clad in green,
The stones and trees insensible of time,
  Nor age nor wrinkle on their front are seen;
If winter come, and greenness then doth fade,
A spring returns, and they 're more youthful made.
But man grows old, lies down, remains where once
    he 's laid.

By birth more noble than those creatures all,
  Yet seems by nature and by custom cursed —
No sooner born but grief and care make fall
  That state obliterate he had at first;
Nor youth, nor strength, nor wisdom spring again,
Nor habitations long their names retain,
But in oblivion to the final day remain.

Shall I then praise the heavens, the trees, the earth,
  Because their beauty and their strength last longer?

Shall I wish there or never to had birth,
    Because they 're bigger and their bodies stronger?
Nay, they shall darken, perish, fade, and die,
And when unmade so ever shall they lie;
But man was made for endless immortality.

Under the cooling shadow of a stately elm
    Close sat I by a goodly river's side
Where gliding streams the rocks did overwhelm;
    A lonely place, with pleasures dignified.
I once that loved the shady woods so well
Now thought the rivers did the trees excel,
And if the sun would ever shine there would I dwell.

While on the stealing stream I fixed mine eye
    Which to the longed-for ocean held its course,
I marked nor crooks nor rubs that there did lie
    Could hinder aught, but still augment its force.
O happy flood, quoth I, that holds thy race
Till thou arrive at thy beloved place,
Nor is it rocks or shoals that can obstruct thy pace.

Nor is it enough that thou alone mayst slide,
    But hundred brooks in thy clear waves do meet;
So hand in hand along with thee they glide
    To Thetis' house, where all embrace and greet.

Thou emblem true of what I count the best,
Oh, could I lead my rivulets to rest!
So may we press to that vast mansion ever blest!

Ye fish which in this liquid region abide,
    That for each season have your habitation,
Now salt, now fresh, where you think best to glide,
    To unknown coasts to give a visitation,
In lakes and ponds you leave your numerous fry;
So nature taught, and yet you know not why,
You watery folk that know not your felicity.

Look how the wantons frisk to taste the air,
    Then to the colder bottom straight they dive;
Eftsoon to Neptune's glassy hall repair
    To see what trade the great ones there do drive;
Who forage o'er the spacious sea-green field,
And take the trembling prey before it yield;
Whose armor is their scales, their spreading fins their
    shield.

While musing thus with contemplation fed,
    And thousand fancies buzzing in my brain,
The sweet-tongued philomel perched o'er my head,
    And chanted forth a most melodious strain,
Which rapt me so with wonder and delight
I judged my hearing better than my sight,
And wished me wings with her a while to take my flight.

O merry bird, said I, that fears no snares,
   That neither toils nor hoards up in thy barn,
Feels no sad thoughts, nor cruciating cares
   To gain more good or shun what might thee harm,
Thy clothes ne'er wear, thy meat is everywhere,
Thy bed a bough, thy drink the water clear,
Reminds not what is past, nor what's to come dost fear!

The dawning morn with songs thou dost prevent,
   Settest hundred notes unto thy feathered crew;
So each one tunes his pretty instrument,
   And, warbling out the old, begins anew.
And thus they pass their youth in summer season,
Then follow thee into a better region
Where winter's never felt by that sweet airy legion.

Man, at the best a creature frail and vain,
   In knowledge ignorant, in strength but weak,
Subject to sorrows, losses, sickness, pain,
   Each storm his state, his mind, his body, break,
From some of these he never finds cessation,
But day or night, within, without, vexation,
Troubles from foes, from friends, from dearest, nearest relation.

And yet this sinful creature, frail and vain,
   This lump of wretchedness, of sin and sorrow,
This weather-beaten vessel racked with pain,
   Joys not in hope of an eternal morrow;

Nor all his losses, crosses, and vexation,
In weight, in frequency, and long duration,
Can make him deeply groan for that divine translation.

The mariner that on smooth waves doth glide
   Sings merrily, and steers his bark with ease,
As if he had command of wind and tide,
   And now become great master of the seas;
But suddenly a storm spoils all the sport,
And makes him long for a more quiet port,
Which 'gainst all adverse winds may serve for fort.

So he that faileth in this world of pleasure,
   Feeding on sweets, that never bit of the sour,
That's full of friends, of honor, and of treasure,
   Fond fool, he takes this earth e'en for heaven's bower.
But sad affliction comes, and makes him see
Here's neither honor, wealth, nor safety;
Only above is found all with security.

O time, the fatal wreck of mortal things,
That draws oblivion's curtains over kings!
Their sumptuous monuments men know them not,
Their names without a record are forgot,
Their parts, their ports, their pomps, all laid in the dust,
Nor wit, nor gold, nor buildings 'scape time's rust.
But he whose name is graved in the white stone
Shall last and shine when all of these are gone.

THE FLESH AND THE SPIRIT.

In secret place where once I stood,
Close by the banks of lacrym flood,
I heard two sisters reason on
Things that are past and things to come.
One Flesh was called, who had her eye
On worldly wealth and vanity;
The other Spirit, who did rear
Her thoughts unto a higher sphere.
"Sister," quoth Flesh, "what livest thou on —
Nothing but meditatiön?
Doth contemplation feed thee, so
Regardlessly to let earth go?
Can speculation satisfy
Notion without reality?
Dost dream of things beyond the moon,
And dost thou hope to dwell there soon?
Hast treasures there laid up in store
That all in the world thou countest poor?
Art fancy sick, or turned a sot,
To catch at shadows which are not?
Come, come, I 'll show unto thy sense
Industry hath its recompense.
What canst desire but thou mayst see
True substance in variety?

Dost honor like? Acquire the same,
As some to their immortal fame,
And trophies to thy name erect
Which wearing time shall ne'er deject.
For riches dost thou long full sore?
Behold enough of precious store;
Earth hath more silver, pearls, and gold
Than eyes can see or hands can hold.
Affectest thou pleasure? Take thy fill;
Earth hath enough of what you will.
Then let not go what thou mayst find
For things unknown, only in mind."

Spirit. "Be still, thou unregenerate part;
Disturb no more my settled heart,
For I have vowed, and so will do,
Thee as a foe still to pursue,
And combat with thee will and must
Until I see thee laid in the dust.
Sisters we are, yea, twins we be,
Yet deadly feud 'twixt thee and me;
For from one father are we not.
Thou by old Adam wast begot,
But my arise is from above,
Whence my dear Father I do love.
Thou speakest me fair, but hatest me sore;
Thy flattering shows I'll trust no more.
How oft thy slave hast thou me made
When I believed what thou hast said,

The Flesh and the Spirit

And never had more cause of woe
Than when I did what thou bad'st do.
I 'll stop mine ears at these thy charms,
And count them for my deadly harms.
Thy sinful pleasures I do hate,
Thy riches are to me no bait,
Thine honors do nor will I love,
For my ambition lies above.
My greatest honor it shall be
When I am victor over thee,
And triumph shall, with laurel head,
When thou my captive shalt be led.
How I do live thou needst not scoff,
For I have meat thou knowest not of:
The hidden manna I do eat,
The word of life it is my meat.
My thoughts do yield me more content
Than can thy hours in pleasure spent.
Nor are they shadows which I catch,
Nor fancies vain at which I snatch,
But reach at things that are so high
Beyond thy dull capacity.
Eternal substance I do see,
With which enrichéd I would be;
Mine eye doth pierce the heavens, and see
What is invisible to thee.
My garments are not silk or gold,
Nor such like trash which earth doth hold,

But royal robes I shall have on,
More glorious than the glistering sun.
My crown not diamonds, pearls, and gold,
But such as angels' heads enfold.
The city where I hope to dwell
There's none on earth can parallel:
The stately walls, both high and strong,
Are made of precious jasper stone;
The gates of pearl both rich and clear,
And angels are for porters there;
The streets thereof transparent gold,
Such as no eye did e'er behold;
A crystal river there doth run,
Which doth proceed from the Lamb's throne;
Of life there are the waters sure,
Which shall remain for ever pure;
Of sun or moon they have no need,
For glory doth from God proceed —
No candle there, nor yet torch-light,
For there shall be no darksome night.
From sickness and infirmity
For evermore they shall be free,
Nor withering age shall e'er come there,
But beauty shall be bright and clear.
This city pure is not for thee,
For things unclean there shall not be.
If I of Heaven may have my fill,
Take thou the world, and all that will.

THE VANITY OF ALL WORLDLY THINGS.

As he said vanity, so vain say I.
O vanity, O vain all under sky!
Where is the man can say, Lo, I have found
On brittle earth a consolation sound?
What, is it in honor to be set on high?
No; they like beasts and sons of men shall die;
And whilst they live, how oft doth turn their fate —
He 's now a captive that was king of late.
What, is it in wealth, great treasures to obtain?
No; that 's but labor, anxious care, and pain.
He heaps up riches, and he heaps up sorrow;
It 's his to-day, but who 's his heir to-morrow?
What, then, content in pleasures canst thou find?
More vain than all, that 's but to grasp the wind.
The sensual senses for a time they please;
Meanwhile the conscience rage, who shall appease?
What, is it in beauty? No; that 's but a snare;
They 're foul enough to-day that once were fair.
What, is it in flowering youth, or manly age?
The first is prone to vice, the last to rage.
Where is it, then, in wisdom, learning, arts?
Sure if on earth it must be in those parts.

Yet these the wisest man of men did find
But vanity, vexation of the mind;
And he that knows the most doth still bemoan
He knows not all that here is to be known.
What is it then, to do as stoics tell —
Nor laugh nor weep, let things go ill or well?
Such stoics are but stocks, such teaching vain;
While man is man, he shall have ease or pain.
If not in honor, beauty, age, or treasure,
Nor yet in learning, wisdom, youth, or pleasure,
Where shall I climb, sound, seek, or search, or find
That *summum bonum* which may stay my mind?
There is a path no vulture's eye hath seen,
Where lion fierce nor lion's whelps have been,
Which leads unto that living crystal fount
Who drinks thereof the world doth naught account.
The depth and sea have said 't is not in me;
With pearl and gold it shall not valued be.
For sapphire, onyx, topaz, who would change?
It 's hid from eyes of men; they count it strange.
Death and destruction the fame hath heard,
But where and what it is from heaven 's declared.
It brings to honor which shall ne'er decay;
It stores with wealth which time can't wear away;
It yieldeth pleasures far beyond conceit,
And truly beautifies without deceit;
Nor strength, nor wisdom, nor fresh youth shall fade,
Nor death shall see, but are immortal made.

This pearl of price, this tree of life, this spring,
Who is possesséd of shall reign a king,
Nor change of state nor cares shall ever see,
But wear his crown unto eternity.
This satiates the soul; this stays the mind;
And all the rest but vanity we find.

THE AUTHOR TO HER BOOK.

Thou ill-formed offspring of my feeble brain,
Who after birth didst by my side remain
Till snatched from thence by friends less wise than true
Who thee abroad exposed to public view,
Made thee, in rags, halting, to the press to trudge,
Where errors were not lessened, all may judge,
At thy return my blushing was not small
My rambling brat — in print — should mother call.
I cast thee by as one unfit for light,
Thy visage was so irksome in my sight;
Yet being mine own, at length affection would
Thy blemishes amend, if so I could.
I washed thy face, but more defects I saw,
And rubbing off a spot still made a flaw.
I stretched thy joints to make thee even feet,
Yet still thou runnest more hobbling than is meet.
In better dress to trim thee was my mind,
But naught save homespun cloth in the house I find.
In this array 'mongst vulgars mayst thou roam,
In critics' hands beware thou dost not come,
And take thy way where yet thou art not known.
If for thy father asked, say thou hadst none;
And for thy mother, she, alas, is poor,
Which caused her thus to send thee out of door.

Several other poems made by the author upon divers occasions were found among her papers after her death, which she never meant should come to public view; amongst which these following, at the desire of some friends that knew her well, are here inserted.

UPON A FIT OF SICKNESS, ANNO 1632.
ÆTATIS SUÆ 19.

Twice ten years old not fully told
  Since nature gave me breath,
My race is run, my thread is spun,
  Lo, here is fatal Death.
All men must die, and so must I,
  This cannot be revoked;
For Adam's sake this word God spake
  When he so high provoked.
Yet live I shall — this life 's but small —
  In place of highest bliss,
Where I shall have all I can crave;
  No life is like to this.
For what 's this life but care and strife?
  Since first we came from womb

Our strength doth waste, our time doth haste,
   And then we go to the tomb.
O bubble blast, how long canst last
   That always art a-breaking? —
No sooner blown but dead and gone,
   E'en as a word that's speaking.
Oh, whilst I live this grace me give,
   I doing good may be,
Then death's arrest I shall count best,
   Because it's thy decree.
Bestow much cost there's nothing lost
   To make salvation sure;
Oh, great's the gain, though got with pain,
   Comes by profession pure.
The race is run, the field is won,
   The victory's mine, I see,
For ever know, thou envious foe,
   The foil belongs to thee.

UPON SOME DISTEMPER OF BODY.

In anguish of my heart replete with woes,
And wasting pains which best my body knows,
In tossing slumbers on my wakeful bed,
Bedrenched with tears that flowed from mournful head
Till nature had exhausted all her store,
Then eyes lay dry, disabled to weep more,

And looking up unto his throne on high
Who sendeth help to those in misery,
He chased away those clouds, and let me see
My anchor cast in the vale with safety;
He eased my soul of woe, my flesh of pain,
And brought me to the shore from troubled main.

BEFORE THE BIRTH OF ONE OF HER CHILDREN.

All things within this fading world have end.
Adversity doth still our joys attend;
No ties so strong, no friends so dear and sweet,
But with death's parting blow are sure to meet.
The sentence passed is most irrevocable,
A common thing, yet, oh, inevitable.
How soon, my dear, death may my steps attend,
How soon it may be thy lot to lose thy friend,
We both are ignorant; yet love bids me
These farewell lines to recommend to thee,
That when that knot's untied that made us one
I may seem thine who in effect am none.
And if I see not half my days that are due,
What nature would God grant to yours and you.
The many faults that well you know I have
Let be interred in my oblivion's grave;
If any worth or virtue were in me,
Let that live freshly in thy memory,

And when thou feelest no grief, as I no harms,
Yet love thy dead, who long lay in thine arms;
And when thy loss shall be repaid with gains
Look to my little babes, my dear remains,
And if thou love thyself, or lovedst me,
These oh protect from stepdam's injury.
And if chance to thine eyes shall bring this verse,
With some sad sighs honor my absent hearse;
And kiss this paper for thy love's dear sake,
Who with salt tears this last farewell did take.

A. B.

TO MY DEAR AND LOVING HUSBAND.

If ever two were one, then surely we;
If ever man were loved by wife, then thee;
If ever wife was happy in a man,
Compare with me, ye women, if you can.
I prize thy love more than whole mines of gold,
Or all the riches that the East doth hold.
My love is such that rivers cannot quench,
Nor aught but love from thee give recompense.
Thy love is such I can no way repay;
The heavens reward thee manifold, I pray.
Then while we live in love let's so persevere
That when we live no more we may live ever.

A LETTER TO HER HUSBAND, ABSENT UPON PUBLIC EMPLOYMENT.

My head, my heart, mine eyes, my life,—nay, more,
My joy, my magazine of earthly store,—
If two be one, as surely thou and I.
How stayest thou there, whilst I at Ipswich lie?—
So many steps head from the heart to sever;
If but a neck soon should we be together.
I, like the earth this season, mourn in black,
My sun is gone so far in his zodiac,
Whom whilst I enjoyed nor storms nor frosts I
    felt,
His warmth such frigid colds did cause to melt.
My chilléd limbs now numbéd lie forlorn;
Return, return, sweet Sol, from Capricorn!
In this dead time, alas, what can I more
Than view those fruits which through thy heat I
    bore?—
Which sweet contentment yield me for a space,
True living pictures of their father's face.
O strange effect! now thou art southward gone
I weary grow, the tedious day so long;
But when thou northward to me shalt return
I wish my sun may never set, but burn
Within the Cancer of my glowing breast,
The welcome house of him my dearest guest,

Where ever, ever stay, and go not thence
Till nature's sad decree shall call thee hence.
Flesh of thy flesh, bone of thy bone,
I here, thou there, yet both but one.

A. B.

ANOTHER.

Phœbus, make haste, the day 's too long, be gone;
The silent night 's the fittest time for moan.
But stay this once, unto my suit give ear,
And tell my griefs in either hemisphere;
And if the whirling of thy wheels don't drown
The woeful accents of my doleful sound,
If in thy swift career thou canst make stay,
I crave this boon, this errand by the way:
Commend me to the man more loved than life;
Show him the sorrows of his widowed wife —
My dumpish thoughts, my groans, my brackish tears,
My sobs, my longing hopes, my doubting fears;
And if he love, how can he there abide?
My interest 's more than all the world beside.
He that can tell the stars or ocean sand,
Or all the grass that in the meads do stand,
The leaves in the woods, the hails, or drops of rain,
Or in a corn-field number every grain,
Or every mote that in the sunshine hops,
May count my sighs, and number all my drops.
Tell him the countless steps that thou dost trace
That once a day thy spouse thou mayst embrace;

And when thou canst not treat by loving mouth
Thy rays afar salute her from the south.
But for one month I see no day, poor soul,
Like those far situate under the Pole,
Which day by day long wait for thy arise;
Oh, how they joy when thou dost light the skies!
O Phœbus, hadst thou but thus long from thine
Restrained the beams of thy beloved shine,
At thy return, if so thou couldst or durst,
Behold a chaos blacker than the first.
Tell him here 's worse than a confuséd matter —
His little world 's a fathom under water;
Naught but the fervor of his ardent beams
Hath power to dry the torrent of these streams.
Tell him I would say more, but cannot well;
Oppresséd minds abruptest tales do tell.
Now post with double speed, mark what I say,
By all our loves conjure him not to stay.

ANOTHER.

As loving hind that, hartless, wants her deer
Scuds through the woods and fern with harkening ear,
Perplexed, in every bush and nook doth pry
Her dearest deer might answer ear or eye,
So doth my anxious soul, which now doth miss
A dearer dear, far dearer heart, than this,
Still wait with doubts, and hopes, and failing eye
His voice to hear or person to descry.

Or as the pensive dove doth all alone
On withered bough most uncouthly bemoan
The absence of her love and loving mate
Whose loss hath made her so unfortunate,
E'en thus do I, with many a deep sad groan,
Bewail my turtle true who now is gone,
His presence and his safe return still woo
With thousand doleful sighs and mournful coo.
Or as the loving mullet, that true fish,
Her fellow lost nor joy nor life doth wish,
But launches on that shore there for to die
Where she her captive husband doth espy,
Mine being gone, I lead a joyless life.
I have a loving feer, yet seem no wife;
But worst of all, to him can't steer my course—
I here, he there, alas, both kept by force.
Return, my dear, my joy, my only love,
Unto thy hind, thy mullet, and thy dove,
Who neither joys in pasture, house, nor streams;
The substance gone, oh me, these are but dreams.
Together at one tree oh let us browse,
And like two turtles roost within one house,
And like the mullets in one river glide—
Let 's still remain but one, till death divide.

    Thy loving love and dearest dear,
      At home, abroad, and everywhere,

                    A. B.

TO HER FATHER, WITH SOME VERSES.

Most truly honored, and as truly dear,
If worth in me or aught I do appear
Who can of right better demand the same
Than may your worthy self, from whom it came?
The principal might yield a greater sum,
Yet, handled ill, amounts but to this crumb.
My stock 's so small I know not how to pay,
My bond remains in force unto this day;
Yet for part payment take this simple mite.
Where nothing 's to be had kings lose their right.
Such is my debt I may not say "Forgive!"
But as I can I 'll pay it while I live;
Such is my bond none can discharge but I,
Yet, paying, is not paid until I die.

A. B.

IN REFERENCE TO HER CHILDREN, 23 JUNE, 1659.

I had eight birds hatched in one nest;
Four cocks there were, and hens the rest.
I nursed them up with pain and care,
Nor cost nor labor did I spare,
Till at the last they felt their wing,
Mounted the trees, and learned to sing.

Chief of the brood then took his flight
To regions far, and left me quite;
My mournful chirps I after send
Till he return or I do end:
Leave not thy nest, thy dam, and sire;
Fly back and sing amidst this choir.
My second bird did take her flight,
And with her mate flew out of sight;
Southward they both their course did bend,
And seasons twain they there did spend,
Till after, blown by southern gales,
They northward steered with filléd sails.
A prettier bird was nowhere seen
Along the beach, among the treen.
I have a third, of color white,
On whom I placed no small delight;
Coupled with mate loving and true,
Hath also bid her dam adieu,
And where Aurora first appears
She now hath perched to spend her years.
One to the academy flew
To chat among that learned crew;
Ambition moves still in his breast
That he might chant above the rest,
Striving for more than to do well —
That nightingales he might excel.
My fifth, whose down is yet scarce gone,
Is 'mongst the shrubs and bushes flown,

In Reference to Her Children

And as his wings increase in strength
On higher boughs he 'll perch at length.
My other three still with me nest
Until they 're grown; then, as the rest,
Or here or there they 'll take their flight;
As is ordained, so shall they light.
If birds could weep, then would my tears
Let others know what are my fears
Lest this my brood some harm should catch
And be surprised for want of watch:
Whilst pecking corn, and void of care,
They fall unawares in fowler's snare;
Or whilst on trees they sit and sing,
Some untoward boy at them do fling;
Or whilst allured with bell and glass,
The net be spread, and caught, alas!
Or lest by lime-twigs they be foiled,
Or by some greedy hawks be spoiled.
Oh, would, my young, ye saw my breast,
And knew what thoughts there sadly rest.
Great was my pain when I you bred,
Great was my care when I you fed;
Long did I keep you soft and warm,
And with my wings kept off all harm.
My cares are more, and fears, than ever,
My throbs such now as 'fore were never.
Alas, my birds, you wisdom want;
Of perils you are ignorant —

Ofttimes in grass, on trees, in flight,
Sore accidents on you may light.
Oh, to your safety have an eye;
So happy may you live and die.
Meanwhile my days in tunes I'll spend
Till my weak lays with me shall end;
In shady woods I'll sit and sing,
Things that are past to mind I'll bring —
Once young and pleasant, as are you.
But former toys,— not joys,— adieu!
My age I will not once lament,
But sing my time so near is spent,
And from the top bough take my flight
Into a country beyond sight,
Where old ones instantly grow young,
And there with seraphims set song.
No seasons cold nor storms they see,
But spring lasts to eternity.
When each of you shall in your nest
Among your young ones take your rest,
In chirping language oft them tell
You had a dam that loved you well,
That did what could be done for young,
And nursed you up till you were strong;
And 'fore she once would let you fly
She showed you joy and misery,
Taught what was good, and what was ill,
What would save life, and what would kill.

Thus gone, amongst you I may live,
And dead, yet speak, and counsel give.
Farewell, my birds, farewell, adieu!
I happy am if well with you.

IN MEMORY OF MY DEAR GRANDCHILD ELIZABETH BRADSTREET, WHO DECEASED AUGUST, 1665, BEING A YEAR AND A HALF OLD.

Farewell, dear babe, my heart's too much content!
   Farewell, sweet babe, the pleasure of mine eye!
Farewell, fair flower that for a space was lent,
   Then taken away unto eternity!
Blest babe, why should I once bewail thy fate,
Or sigh the days so soon were terminate,
Since thou art settled in an everlasting state?

By nature trees do rot when they are grown,
   And plums and apples throughly ripe do fall,
And corn and grass are in their season mown,
   And time brings down what is both strong and tall.
But plants new set to be eradicate,
And buds new blown to have so short a date,
Is by His hand alone that guides nature and fate.

IN MEMORY OF MY DEAR GRANDCHILD ANNE BRADSTREET, WHO DECEASED JUNE 20, 1669, BEING THREE YEARS AND SEVEN MONTHS OLD.

With troubled heart and trembling hand I write.
The heavens have changed to sorrow my delight.
How oft with disappointment have I met
When I on fading things my hopes have set.
Experience might 'fore this have made me wise
To value things according to their price.
Was ever stable joy yet found below?
Or perfect bliss without mixture of woe?
I knew she was but as a withering flower,
That 's here to-day, perhaps gone in an hour;
Like as a bubble, or the brittle glass,
Or like a shadow turning, as it was.
More fool, then, I to look on that was lent
As if mine own, when thus impermanent.
Farewell, dear child; thou ne'er shalt come to me,
But yet a while and I shall go to thee.
Meantime my throbbing heart 's cheered up with this—
Thou with thy Saviour art in endless bliss.

ON MY DEAR GRANDCHILD SIMON BRADSTREET, WHO DIED ON 16TH NOVEMBER, 1669, BEING BUT A MONTH AND ONE DAY OLD.

No sooner come but gone, and fallen asleep;
Acquaintance short, yet parting caused us weep.
Three flowers—two scarcely blown, the last in bud—
Cropped by the Almighty's hand! Yet is he good.
With dreadful awe before him let's be mute.
Such was his will, but why let's not dispute.
With humble hearts and mouths put in the dust
Let's say he's merciful as well as just.
He will return, and make up all our losses,
And smile again, after our bitter crosses.
Go, pretty babe; go rest with sisters twain;
Among the blest in endless joys remain.

TO THE MEMORY OF MY DEAR DAUGHTER-IN-LAW MRS. MERCY BRADSTREET, WHO DECEASED SEPTEMBER 6, 1670, IN THE 28TH YEAR OF HER AGE.

And live I still to see relations gone?
And yet survive to sound this wailing tone?
Ah, woe is me, to write thy funeral song
Who might in reason yet have livéd long.

I saw the branches lopped, the tree now fall,
I stood so nigh it crushed me down withal;
My bruiséd heart lies sobbing at the root
That thou, dear son, hath lost both tree and fruit.
Thou then, on seas sailing to foreign coast,
Wast ignorant what riches thou hadst lost;
But, ah! too soon those heavy tidings fly
To strike thee with amazing misery.
Oh, how I sympathize with thy sad heart,
And in thy griefs still bear a second part.
I lost a daughter dear, but thou a wife
Who loved thee more, it seemed, than her own life —
Thou being gone, she longer could not be
Because her soul she 'd sent along with thee.
One week she only passed in pain and woe,
And then her sorrows all at once did go.
A babe she left before she soared above,
The fifth and last pledge of her dying love.
Ere nature would it hither did arrive;
No wonder it no longer did survive.
So with her children four she 's now at rest,
All freed from grief, I trust, among the blest.
She one hath left, a joy to thee and me;
The heavens vouchsafe she may so ever be.
Cheer up, dear son, thy fainting bleeding heart
In Him alone that causéd all this smart.
What though thy strokes full sad and grievous be?
He knows it is the best for thee and me.

A FUNERAL ELEGY UPON THAT PATTERN AND PATRON OF VIRTUE, THE TRULY PIOUS, PEERLESS, AND MATCHLESS GENTLEWOMAN MRS. ANNE BRADSTREET, RIGHT PANARETES, MIRROR OF HER AGE, GLORY OF HER SEX, WHOSE HEAVEN-BORN SOUL, LEAVING ITS EARTHLY SHRINE, CHOSE ITS NATIVE HOME AND WAS TAKEN TO ITS REST UPON 16TH SEPTEMBER, 1672.

Ask not why hearts turn magazines of passions,
And why that grief is clad in several fashions;
Why she on progress goes, and doth not borrow
The smallest respite from the extremes of sorrow.
Her misery is got to such an height
As makes the earth groan to support its weight;
Such storms of woe so strongly have beset her
She hath no place for worse nor hope for better.
Her comfort is, if any for her be,
That none can show more cause of grief than she.

Ask not why some in mournful black are clad:
The sun is set; there needs must be a shade.
Ask not why every face a sadness shrouds:
The setting sun o'ercast us hath with clouds.
Ask not why the great glory of the sky,
That gilds the stars with heavenly alchemy,
Which all the world doth lighten with his rays,
The Persian god, the monarch of the days —
Ask not the reason of his ecstasy,
Paleness of late, in midnoon majesty;
Why that the palefaced empress of the night
Disrobed her brother of his glorious light.
Did not the language of the stars foretell
A mournful scene when they with tears did swell?
Did not the glorious people of the sky
Seem sensible of future misery?
Did not the lowering heavens seem to express
The world's great loss, and their unhappiness?
Behold how tears flow from the learned hill,
How the bereavéd Nine do daily fill
The bosom of the fleeting air with groans
And woeful accents, which witness their moans;
How do the goddesses of verse, the learned choir,
Lament their rival quill, which all admire.
Could Maro's muse but hear her lively strain
He would condemn his works to fire again.
Methinks I hear the patron of the spring,
The unshorn deity, abruptly sing:

A Funeral Elegy upon Mrs. Anne Bradstreet

"Some do for anguish weep; for anger I
That ignorance should live and art should die.
Black, fatal, dismal, inauspicious day,
Unblest for ever by Sol's precious ray,
Be it the first of miseries to all,
Or last of life, defamed for funeral.
When this day yearly comes let every one
Cast in their urn the black and dismal stone;
Succeeding years as they their circuit go
Leap o'er this day, as a sad time of woe.
Farewell, my muse; since thou hast left thy shrine
I am unblest in One, but blest in Nine.
Fair Thespian ladies, light your torches all;
Attend your glory to its funeral.
To court her ashes with a learned tear,
A briny sacrifice, let not a smile appear."
Grave matron, whoso seeks to blazon thee
Needs not make use of wit's false heraldry;
Whoso should give thee all thy worth would swell
So high as it would turn the world infidel.
Had he great Maro's muse, or Tully's tongue,
Or raping numbers like the Thracian song,
In crowning of her merits he would be
Sumptuously poor, low in hyperbole.
To write is easy; but to write on thee
Truth would be thought to forfeit modesty.
He'll seem a poet that shall speak but true;
Hyperboles in others are thy due.

Like a most servile flatterer he will show,
Though he write truth, and make the subject you.
Virtue ne'er dies; time will a poet raise,
Born under better stars, shall sing thy praise.
Praise her who list, yet he shall be a debtor,
For art ne'er feigned nor nature framed a better.
Her virtues were so great that they do raise
A work to trouble fame, astonish praise.
Whenas her name doth but salute the ear,
Men think that they perfection's abstract hear.
Her breast was a brave palace, a Broad-street,
Where all heroic ample thoughts did meet,
Where nature such a tenement had ta'en
That others' souls to hers dwelt in a lane.
Beneath her feet pale envy bites her chain,
And poison malice whets her sting in vain.
Let every laurel, every myrtle bough,
Be stripped for leaves to adorn and load her
    brow —
Victorious wreaths, which 'cause they never fade
Wise elder times for kings and poets made.
Let not her happy memory e'er lack
Its worth in fame's eternal almanac,
Which none shall read but straight their loss
    deplore,
And blame their fates they were not born before.
Do not old men rejoice their fates did last,
And infants, too, that theirs did make such haste

In such a welcome time to bring them forth
That they might be a witness to her worth?
Who undertakes this subject to commend
Shall nothing find so hard as how to end.

> *Finis & Non.*
>
> JOHN NORTON.

Omnia Romanæ fileant miracula gentis.

[All the following compositions of Mrs. Bradstreet were first printed in the 1867 edition of her writings edited by Mr. John H. Ellis. He found them in a small journal in the possession of one of her descendants. The "Meditations Divine and Moral" were in the handwriting of Mrs. Bradstreet herself.]

FOR MY DEAR SON SIMON BRADSTREET.

Parents perpetuate their lives in their posterity, and their manners in their imitation. Children do naturally rather follow the failings than the virtues of their predecessors; but I am persuaded better things of you. You once desired me to leave something for you in writing that you might look upon when you should see me no more. I could think of nothing more fit for you, nor of more ease to myself, than these short meditations following. Such as they are I bequeath to you: small legacies are accepted by true friends, much more by dutiful children. I have avoided encroaching upon others' conceptions, because I would leave you nothing but mine own; though in value they fall short of all in this kind, yet I presume they will be better prized by you for the author's sake. The Lord bless you with grace here, and crown you with glory hereafter, that I may meet you with rejoicing at that great day of appearing, which is the continual prayer of

Your affectionate mother,

March 20, 1664. A. B.

MEDITATIONS DIVINE AND MORAL.

There is no object that we see, no action that we do, no good that we enjoy, no evil that we feel or fear, but we may make some spiritual advantage of all; and he that makes such improvement is wise as well as pious.

Many can speak well, but few can do well. We are better scholars in the theory than the practice part; but he is a true Christian that is a proficient in both.

Youth is the time of getting, middle age of improving, and old age of spending; a negligent youth is usually attended by an ignorant middle age, and both by an empty old age. He that hath nothing to feed on but vanity and lies must needs lie down in the bed of sorrow.

A ship that bears much sail, and little or no ballast, is easily overset; and that man whose head hath great abilities, and his heart little or no grace, is in danger of foundering.

It is reported of the peacock that priding himself in his gay feathers he ruffles them up; but spying his black feet he soon lets fall his plumes. So he that glories in his gifts and adornings should look upon his corruptions, and that will damp his high thoughts.

Meditations Divine and Moral

The finest bread hath the least bran, the purest honey the least wax, and the sincerest Christian the least self-love.

The hireling that labors all the day comforts himself that when night comes he shall both take his rest and receive his reward. The painful Christian that hath wrought hard in God's vineyard, and hath borne the heat and drought of the day, when he perceives his sun apace to decline, and the shadows of his evening to be stretched out, lifts up his head with joy, knowing his refreshing is at hand.

Downy beds make drowsy persons, but hard lodging keeps the eyes open. A prosperous state makes a secure Christian, but adversity makes him consider.

Sweet words are like honey: a little may refresh, but too much gluts the stomach.

Diverse children have their different natures: some are like flesh which nothing but salt will keep from putrefaction; some again like tender fruits that are best preserved with sugar. Those parents are wise that can fit their nurture according to their nature.

That town which thousands of enemies without hath not been able to take hath been delivered up by one traitor within; and that man which all the temptations of Satan without could not hurt hath been soiled by one lust within.

Authority without wisdom is like a heavy axe without an edge — fitter to bruise than polish.

The reason why Christians are so loth to exchange this world for a better is because they have more sense than faith: they see what they enjoy, they do but hope for that which is to come.

If we had no winter, the spring would not be so pleasant; if we did not sometimes taste of adversity, prosperity would not be so welcome.

A low man can go upright under that door where a taller is glad to stoop; so a man of weak faith and mean abilities may undergo a cross more patiently than he that excels him both in gifts and graces.

That house which is not often swept makes the cleanly inhabitant soon loathe it; and that heart which is not continually purifying itself is no fit temple for the spirit of God to dwell in.

Few men are so humble as not to be proud of their abilities; and nothing will abase them more than this: What hast thou but what thou hast received? Come, give an account of thy stewardship.

He that will undertake to climb up a steep mountain with a great burden on his back will find it a wearisome if not an impossible task; so he that thinks to mount to heaven clogged with the cares and riches of this life, 't is no wonder if he faint by the way.

Corn till it has passed through the mill and been ground to powder is not fit for bread. God so deals with his servants: he grinds them with grief and pain till they turn to dust, and then are they fit manchet for his mansion.

God hath suitable comforts and supports for his children according to their several conditions if he will make his face to shine upon them. He then makes them lie down in green pastures, and leads them beside the still waters; if they stick in deep mire and clay, and all his waves and billows go over their heads, he then leads them to the rock which is higher than they.

He that walks among briers and thorns will be very careful where he sets his foot; and he that passes through the wilderness of this world had need ponder all his steps.

Want of prudence as well as piety hath brought men into great inconveniences; but he that is well stored with both seldom is so ensnared.

The skilful fisher hath his several baits for several fish, but there is a hook under all; Satan, that great angler, hath his sundry baits for sundry tempers of men, which they all catch greedily at, but few perceive the hook till it be too late.

There is no new thing under the sun; there is nothing that can be said or done but either that or something like it hath been both done and said before.

An aching head requires a soft pillow, and a drooping heart a strong support.

A sore finger may disquiet the whole body, but an ulcer within destroys it; so an enemy without may disturb a commonwealth, but dissensions within overthrow it.

It is a pleasant thing to behold the light, but sore eyes are not able to look upon it; the pure in heart shall see God, but the defiled in conscience shall rather choose to be buried under rocks and mountains than to behold the presence of the Lamb.

Wisdom with an inheritance is good, but wisdom without an inheritance is better than an inheritance without wisdom.

Lightning doth usually precede thunder, and storms rain; and strokes do not often fall till after threatening.

Yellow leaves argue the want of sap, and gray hairs the want of moisture; so dry and sapless performances are symptoms of little spiritual vigor.

Iron till it be thoroughly heated is incapable to be wrought; so God sees good to cast some men into the furnace of affliction, and then beats them on his anvil into what frame he pleases.

Ambitious men are like hops, that never rest climbing so long as they have anything to stay upon; but take away their props and they are of all the most dejected.

Much labor wearies the body, and many thoughts oppress the mind; man aims at profit by the one and content in the other, but often misses of both, and finds nothing but vanity and vexation of spirit.

Dim eyes are the concomitants of old age; and short-sightedness, in those that are eyes of a republic, foretells a declining state.

We read in Scripture of three sorts of arrows — the arrow of an enemy, the arrow of pestilence, and the arrow of a slanderous tongue. The first two kill the body, the last the good name; the former two leave a man when he is once dead, but the last mangles him in his grave.

Sore laborers have hard hands, and old sinners have brawny consciences.

Wickedness comes to its height by degrees. He that dares say of a less sin, Is it not a little one? will ere long say of a greater, Tush, God regards it not!

Some children are hardly weaned; although the teat be rubbed with wormwood or mustard, they will either wipe it off, or else suck down sweet and bitter together. So is it with some Christians: let God embitter all the sweets of this life, that so they might feed upon more substantial food, yet they are so childishly sottish that they are still hugging and sucking these empty breasts, that God is forced to hedge up their way with thorns, or lay affliction on their loins, that so they might shake hands with the world before it bid them farewell.

A prudent mother will not clothe her little child with a long and cumbersome garment; she easily foresees what events it is like to produce — at the best but falls and bruises, or perhaps somewhat worse. Much more will the All-wise God proportion his dispensations according to the stature and strength of the person he bestows them on. Large endowments of honor, wealth, or a healthful body would quite overthrow some weak Christian; therefore God cuts their garments short, to keep them in such a trim that they might run the ways of his commandment.

The spring is a lively emblem of the Resurrection. After a long winter we see the leafless trees and dry stalks at the approach of the sun to resume their former vigor and beauty in a more ample manner than what they lost in the autumn. So shall it be at that great day, after a long vacation, when the Sun of Righteousness shall appear: those dry bones shall arise in far more glory than that which they lost at their creation, and in this transcend the spring — that their leaf shall never fail, nor their sap decline.

A wise father will not lay a burden on a child of seven years old which he knows is enough for one of twice his strength; much less will our heavenly Father, who knows our mold, lay such afflictions upon his weak children as would crush them to the dust, but according to the strength he will proportion the load. As God hath his little children, so he hath his strong men, such as are come to a full stature in Christ; and many times he imposes weighty burdens on their shoulders, and yet they go upright under them. But it matters not whether the load be more or less if God afford his help.

"I have seen an end of all perfection," said the royal prophet; but he never said, "I have seen an end of all sinning." What he did say may be easily said by many; but what he did not say cannot truly be uttered by any.

Fire hath its force abated by water, not by wind; and anger must be allayed by cold words, and not by blustering threats.

A sharp appetite and a thorough concoction are a sign of an healthful body; so a quick reception and a deliberate cogitation argue a sound mind.

We often see stones hang with drops, not from any innate moisture, but from a thick air about them; so may we sometimes see marble-hearted sinners seem full of contrition, but it is not from any dew of grace within, but from some black clouds that impend them, which produce these sweating effects.

The words of the wise, saith Solomon, are as nails and as goads, both used for contrary ends — the one holds fast, the other puts forward. Such should be the precepts of the wise masters of assemblies to their hearers, not only to bid them hold fast the form of sound doctrine, but also so to run that they might obtain.

A shadow in the parching sun and a shelter in a blustering storm are of all seasons the most welcome; so a faithful friend in time of adversity is of all other most comfortable.

There is nothing admits of more admiration than God's various dispensation of his gifts among the sons of men, betwixt whom he hath put so vast a disproportion that they scarcely seem made of the same lump or sprung out of the loins of one Adam: some set in the highest dignity that mortality is capable of, and some again so base that they are viler than the earth; some so wise and learned that they seem like angels among men, and some again so ignorant and sottish that they are more like beasts than men; some pious saints, some incarnate devils; some exceeding beautiful, and some extremely deformed; some so strong and healthful that their bones are full of marrow and their breasts of milk, and some again so weak and feeble that, while they live, they are accounted among the dead. And no other reason can be given of all this but so it pleased Him whose will is the perfect rule of righteousness.

The treasures of this world may well be compared to husks; for they have no kernel in them, and they that feed upon them may soon stuff their throats but cannot fill their bellies — they may be choked by them, but cannot be satisfied with them.

Sometimes the sun is only shadowed by a cloud that we cannot see his luster, although we may walk by his light; but when he is set we are in darkness till he arise again. So God doth sometimes veil his face but for a moment that we cannot behold the light of his countenance as at some other time; yet he affords so much light as may direct our way, that we may go forward to the city of habitation. But when he seems to set and be quite gone out of sight, then must we needs walk in darkness and see no light; yet then must we trust in the Lord, and stay upon our God, and when the morning, which is the appointed time, is come the Sun of Righteousness will arise with healing in his wings.

The eyes and the ears are the inlets or doors of the soul, through which innumerable objects enter; yet is not that spacious room filled, neither doth it ever say, "It is enough!" but like the daughters of the horse-

leech cries, "Give! Give!" And, which is most strange, the more it receives, the more empty it finds itself, and sees an impossibility ever to be filled but by Him in whom all fulness dwells.

Had not the wisest of men taught us this lesson, that all is vanity and vexation of spirit, yet our own experience would soon have spelled it out; for what do we obtain of all these things but it is with labor and vexation? When we enjoy them it is with vanity and vexation; and if we lose them then they are less than vanity and more than vexation. So that we have good cause often to repeat that sentence, "Vanity of vanities, vanity of vanities; all is vanity."

He that is to sail into a far country, although the ship, cabin, and provision be all convenient and comfortable for him, yet he hath no desire to make that his place of residence, but longs to put in at that port where his business lies. A Christian is sailing through this world unto his heavenly country, and here he hath many conveniences and comforts; but he must beware of desiring to make this the place of his abode, lest he meet with such tossings that may cause him to long for shore before he sees land. We must, therefore, be here as strangers and pilgrims, that we may plainly declare that we seek a city above, and wait all the days of our appointed time till our change shall come.

He that never felt what it was to be sick or

wounded doth not much care for the company of the physician or chirurgeon; but if he perceive a malady that threatens him with death he will gladly entertain him whom he slighted before. So he that never felt the sickness of sin nor the wounds of a guilty conscience cares not how far he keeps from him that hath skill to cure it; but when he finds his diseases to distress him, and that he must needs perish if he have no remedy, will unfeignedly bid him welcome that brings a plaster for his sore or a cordial for his fainting.

We read of ten lepers that were cleansed, but of one that returned thanks. We are more ready to receive mercies than we are to acknowledge them. Men can use great importunity when they are in distresses, and show great ingratitude after their successes; but he that ordereth his conversation aright will glorify him that heard him in the day of his trouble.

The remembrance of former deliverances is a great support in present distresses. "He that delivered me," saith David, "from the paw of the lion and the paw of the bear will deliver me from this uncircumcised Philistine"; and "He that hath delivered me," saith Paul, "will deliver me." God is the same yesterday, to-day, and for ever; we are the same that stand in need of him, to-day as well as yesterday, and so shall for ever.

Great receipts call for great returns; the more that

any man is intrusted withal, the larger his accounts stand upon God's score. It therefore behooves every man so to improve his talents that when his great Master shall call him to reckoning he may receive his own with advantage.

Sin and shame ever go together; he that would be freed from the last must be sure to shun the company of the first.

God doth many times both reward and punish for one and the same action. As we see in Jehu, he is rewarded with a kingdom to the fourth generation for taking vengeance on the house of Ahab; and yet "A little while," saith God, "and I will avenge the blood of Jezebel upon the house of Jehu." He was rewarded for the matter, and yet punished for the manner; which should warn him that doth any special service for God to fix his eye on the command, and not on his own ends, lest he meet with Jehu's reward, which will end in punishment.

He that would be content with a mean condition must not cast his eye upon one that is in a far better estate than himself, but let him look upon him that is lower than he is, and, if he see that such a one bears poverty comfortably, it will help to quiet him; but if that will not do, let him look on his own unworthiness, and that will make him say with Jacob, "I am less than the least of thy mercies."

Corn is produced with much labor, as the husbandman well knows, and some land asks much more pains than some other doth to be brought into tilth; yet all must be plowed and harrowed. Some children, like sour land, are of so tough and morose a disposition that the plow of correction must make long furrows on their back, and the harrow of discipline go often over them, before they be fit soil to sow the seed of morality, much less of grace, in them. But when by prudent nurture they are brought into a fit capacity, let the seed of good instruction and exhortation be sown in the spring of their youth, and a plentiful crop may be expected in the harvest of their years.

As man is called the little world, so his heart may be called the little commonwealth; his more fixed and resolved thoughts are like to inhabitants, his slight and flitting thoughts are like passengers that travel to and fro continually. Here is also the great court of justice erected, which is always kept by conscience, who is both accuser, excuser, witness, and judge, whom no bribes can pervert nor flattery cause to favor, but as he finds the evidence so he absolves or condemns; yea, so absolute is this court of judicature that there is no appeal from it — no, not to the court of Heaven itself. For if our conscience condemn us, he also who is greater than our conscience will do it much more; but he that would have boldness to go to the throne of grace to be accepted there must be sure to carry a certificate from the court of conscience that he stands right there.

He that would keep a pure heart and lead a blameless life must set himself always in the awful presence of God; the consideration of his all-seeing eye will be a bridle to restrain from evil and a spur to quicken on to good duties. We certainly dream of some remoteness betwixt God and us, or else we should not so often fail in our whole course of life as we do; but he that with David sets the Lord always in his sight will not sin against him.

We see in orchards some trees so fruitful that the weight of their burden is the breaking of their limbs; some again are but meanly laden, and some have nothing to show but leaves only, and some among them are dry stalks. So is it in the church, which is God's orchard: there are some eminent Christians that are so frequent in good duties that many times the weight thereof impairs both their bodies and estates; and there are some, and they sincere ones, too, who have not attained to that fruitfulness, although they aim at perfection; and again there are others that have nothing to commend them but only a gay profession, and these are but leafy Christians which are in as much danger of being cut down as the dry stalks, for both cumber the ground.

We see in the firmament there is but one sun among a multitude of stars, and those stars also to differ much one from the other in regard of bigness and brightness; yet all receive their light from that one sun. So is it in the church both militant and triumphant: there is but one Christ, who is the Sun of Righteousness, in the midst of an innumerable company of saints and angels. Those saints have their degrees even in this life: some are stars of the first magnitude, and some of a less degree, and others — and they indeed the most in number — but small and obscure; yet all receive their luster, be it more or less, from that glorious Sun that enlightens all in all. And if some of them shine so bright while they move on earth, how transcendently splendid shall they be when they are fixed in their heavenly spheres!

Men that have walked very extravagantly, and at last bethink themselves of turning to God, the first thing which they eye is how to reform their ways rather than to beg forgiveness for their sins. Nature looks more at a compensation than at a pardon; but he that will not come for mercy without money and without price, but brings his filthy rags to barter for it, shall meet with miserable disappointment, going away empty, bearing the reproach of his pride and folly.

All the works and doings of God are wonderful, but none more awful than his great work of election and reprobation. When we consider how many good parents have had bad children, and again how many bad parents have had pious children, it should make us adore the sovereignty of God, who will not be tied to time nor place, nor yet to persons, but takes and chooses when and where and whom he pleases. It should also teach the children of godly parents to walk with fear and trembling, lest they, through unbelief, fall short of a promise. It may also be a support to such as have or had wicked parents, that if they abide not in unbelief God is able to graft them in. The upshot of all should make us, with the apostle, to admire the justice and mercy of God, and say, How unsearchable are his ways, and his footsteps past finding out.

The gifts that God bestows on the sons of men are not only abused, but most commonly employed for a clean contrary end than that which they were given for — as health, wealth, and honor, which might be so many steps to draw men to God in consideration of his bounty towards them, but have driven them the further from him, that they are ready to say, We are lords; we will come no more at thee. If outward blessings be not as wings to help us mount upwards they will certainly prove clogs and weights that will pull us lower downward.

All the comforts of this life may be compared to the

Meditations Divine and Moral

gourd of Jonah, that notwithstanding we take great delight for a season in them, and find their shadow very comfortable, yet there is some worm or other of discontent, of fear, or grief that lies at the root, which in great part withers the pleasure which else we should take in them; and well it is that we perceive a decay in their greenness, for were earthly comforts permanent, who would look for heavenly?

All men are truly said to be tenants at will, and it may as truly be said that all have a lease of their lives, some longer, some shorter, as it pleases our great Landlord to let. All have their bounds set, over which they cannot pass, and till the expiration of that time no dangers, no sickness, no pains, or troubles shall put a period to our days; the certainty that that time will come, together with the uncertainty how, where, and when, should make us so to number our days as to apply our hearts to wisdom, that when we are put out of these houses of clay we may be sure of an everlasting habitation that fades not away.

All weak and diseased bodies have hourly mementos of their mortality. But the soundest of men have likewise their nightly monitor by the emblem of death, which is their sleep, for so is death often called; and not only their death, but their grave is lively represented before their eyes by beholding their bed. The morning may mind them of the Resurrection;

and the sun, approaching, of the appearing of the Sun of Righteousness, at whose coming they shall all rise out of their beds, the long night shall flee away, and the day of eternity shall never end. Seeing these things must be, what manner of persons ought we to be in all good conversation?

As the brands of a fire, if once severed, will of themselves go out, although you use no other means to extinguish them, so distance of place, together with length of time, if there be no intercourse, will cool the affections of intimate friends, though there should be no displeasance between them.

A good name is as a precious ointment, and it is a great favor to have a good repute among good men. Yet it is not that which commends us to God, for by his balance we must be weighed, and by his judgment we must be tried; and as he passes the sentence, so shall we stand.

Well doth the apostle call riches deceitful riches, and they may truly be compared to deceitful friends who speak fair and promise much, but perform nothing, and so leave those in the lurch that most relied on them. So is it with the wealth, honors, and pleasures of this world, which miserably delude men and make them put great confidence in them; but when death threatens, and distress lays hold upon them, they prove like the reeds of Egypt that pierce instead of supporting, like empty wells in the time of drought, that those that go to find water in them return with their empty pitchers ashamed.

It is admirable to consider the power of faith, by which all things are almost possible to be done. It can remove mountains, if need were; it hath stayed the course of the sun, raised the dead, cast out devils, reversed the order of nature, quenched the violence of the fire, made the water become firm footing for Peter to walk on. Nay, more than all these, it hath overcome the Omnipotent himself, as, when Moses interceded for the people, God said to him, "Let me alone that I may destroy them!"—as if Moses had been able, by the hand of faith, to hold the everlasting arms of the mighty God of Jacob. Yea, Jacob himself, when he wrestled with God face to face in Peniel, "Let me go," said that angel. "I will not let thee go," replied Jacob, "till thou bless me!" Faith is not only thus potent, but it is so necessary that without faith there is no salvation; therefore, with all our seekings and gettings, let us above all seek to obtain this pearl of price.

Some Christians do by their lusts and corruptions as the Israelites did by the Canaanites, not destroy them, but put them under tribute; for that they could do, as they thought, with less hazard and more profit. But what was the issue? They became a snare unto them, pricks in their eyes and thorns in their sides, and at last overcame them and kept them under slavery. So it is most certain that those that are disobedient to the command of God, and endeavor not to the utmost to drive out all their accursed inmates, but make a league with them, they shall at last fall into perpetual bondage under them unless the great deliverer, Christ Jesus, come to their rescue.

God hath by his providence so ordered that no one country hath all commodities within itself, but what it wants another shall supply, that so there may be a mutual commerce through the world. As it is with countries so it is with men: there was never yet any one man that had all excellences; let his parts, natural and acquired, spiritual and moral, be never so large, yet he stands in need of something which another man hath, perhaps meaner than himself, which shows us perfection is not below, as also that God will have us beholden one to another.

["My honored and dear mother intended to have filled up this book with the like observations, but was prevented by death."—Note by Simon Bradstreet, Jr.]

[The matter on the succeeding pages was at a later date copied into the same journal by her son Simon, with this note: "A true copy of a book left by my honored and dear mother to her children, and found among some papers after her death."]

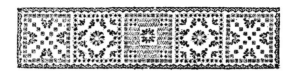

TO MY DEAR CHILDREN.

This book, by any yet unread,
I leave for you when I am dead,
That, being gone, here you may find
What was your living mother's mind.
Make use of what I leave in love,
And God shall bless you from above.

A. B.

My dear Children:

I, knowing by experience that the exhortations of parents take most effect when the speakers leave to speak, and those especially sink deepest which are spoke latest, and being ignorant whether on my death-bed I shall have opportunity to speak to any of you, much less to all, thought it the best, whilst I was able, to compose some short matters (for what else to call them I know not) and bequeath to you, that when I am no more with you yet I may be daily in your remembrance — although that is the least in my aim in what I now do, but that you may gain some spiritual advantage by my experience. I have not studied in this you read to show my skill, but to declare the truth; not to set forth myself, but the glory of God. If I had minded the former, it had been perhaps better pleasing to you; but seeing the last is the best, let it be best pleasing to you.

The method I will observe shall be this: I will begin with God's dealing with me from my childhood to this day. In my young years, about six or seven as I take it, I began to make conscience of my ways, and what I knew was sinful — as lying, disobedience to parents, etc. — I avoided it. If at any time I was overtaken with the like evils, it was a great trouble. I could not be at rest till by prayer I had confessed it unto God. I was also troubled at the neglect of private duties, though too often tardy that way. I also found much comfort in reading the Scriptures, especially those places I thought most concerned my condition; and as I grew to have more understanding, so the more solace I took in them.

In a long fit of sickness which I had on my bed I often communed with my heart, and made my supplication to the Most High, who set me free from that affliction.

But as I grew up to be about fourteen or fifteen I found my heart more carnal, and, sitting loose from God, vanity and the follies of youth take hold of me.

About sixteen the Lord laid his hand sore upon me and smote me with the small-pox. When I was in

To My Dear Children

my affliction, I besought the Lord, and confessed my pride and vanity, and he was entreated of me and again restored me. But I rendered not to him according to the benefit received.

After a short time I changed my condition and was married, and came into this country, where I found a new world and new manners, at which my heart rose. But after I was convinced it was the way of God, I submitted to it and joined to the church at Boston.

After some time I fell into a lingering sickness like a consumption, together with a lameness, which correction I saw the Lord sent to humble and try me and do me good; and it was not altogether ineffectual.

It pleased God to keep me a long time without a child, which was a great grief to me, and cost me many prayers and tears before I obtained one; and after him gave me many more, of whom I now take the care, that as I have brought you into the world, and with great pains, weakness, cares, and fears brought you to this, I now travail in birth again of you till Christ be formed in you.

Among all my experiences of God's gracious dealings with me I have constantly observed this, that he hath never suffered me long to sit loose from him, but by one affliction or other hath made me look home, and search what was amiss; so usually thus it hath been with me that I have no sooner felt my heart out of order but I have expected correction for it, which most commonly hath been upon my own person in sickness, weakness, pains, sometimes on my soul in doubts and fears of God's displeasure and my sincerity towards him. Sometimes he hath smote a child with sickness, sometimes chastened by losses in estate; and these times, through his great mercy, have been the times of my greatest getting and advantage — yea, I have found them the times when the Lord hath manifested the most love to me. Then have I gone to searching, and have said with David, "Lord, search me and try me, see what ways of wickedness are in me, and lead me in the way everlasting." And seldom or never but I have found either some sin I lay under which God would have reformed, or some duty neglected which he would have performed. And by his help I have laid vows and bonds upon my soul to perform his righteous commands.

If at any time you are chastened of God, take it as thankfully and joyfully as in greatest mercies; for if ye be his ye shall reap the greatest benefit by it. It hath been no small support to me in times of darkness when the Almighty hath hid his face from me that yet I have had abundance of sweetness and refreshment after affliction, and more circumspection in my walking after I have been afflicted. I have been with God like an untoward child, that no longer than the rod has been on my back, or at least in sight, but I have been apt to forget him and myself too. "Before

To My Dear Children

I was afflicted I went astray, but now I keep thy statutes."

I have had great experience of God's hearing my prayers and returning comfortable answers to me, either in granting the thing I prayed for or else in satisfying my mind without it; and I have been confident it hath been from him, because I have found my heart through his goodness enlarged in thankfulness to him.

I have often been perplexed that I have not found that constant joy in my pilgrimage and refreshing which I supposed most of the servants of God have; although he hath not left me altogether without the witness of his Holy Spirit, who hath oft given me his word and set to his seal that it shall be well with me. I have sometimes tasted of that hidden manna that the world knows not, and have set up my Ebenezer, and have resolved with myself that against such a promise, such tastes of sweetness, the gates of hell shall never prevail. Yet have I many times sinkings and droopings, and not enjoyed that felicity that sometimes I have done. But when I have been in darkness, and seen no light, yet have I desired to stay myself upon the Lord.

And when I have been in sickness and pain I have thought if the Lord would but lift up the light of his countenance upon me, although he ground me to powder it would be but light to me; yea, oft have

I thought, were it hell itself, and could there find the love of God toward me, it would be a heaven. And could I have been in heaven without the love of God, it would have been a hell to me; for, in truth, it is the absence and presence of God that makes heaven or hell.

Many times hath Satan troubled me concerning the verity of the Scriptures; many times by atheism how I could know whether there was a God. I never saw any miracles to confirm me, and those which I read of how did I know but they were feigned? That there is a God my reason would soon tell me by the wondrous works that I see — the vast frame of the heaven and the earth, the order of all things, night and day, summer and winter, spring and autumn, the daily providing for this great household upon the earth, the preserving and directing of all to its proper end. The consideration of these things would with amazement certainly resolve me that there is an Eternal Being.

But how should I know he is such a God as I worship in Trinity, and such a Saviour as I rely upon? Though this hath thousands of times been suggested to me, yet God hath helped me over. I have argued thus with myself: That there is a God I see. If ever this God hath revealed himself, it must be in his Word, and this must be it or none. Have I not found that operation by it that no human invention can

To My Dear Children

work upon the soul? Have not judgments befallen divers who have scorned and contemned it? Hath it not been preserved through all ages maugre all the heathen tyrants and all of the enemies who have opposed it? Is there any story but that which shows the beginnings of time, and how the world came to be as we see? Do we not know the prophecies in it fulfilled which could not have been so long foretold by any but God himself?

When I have got over this block then have I another put in my way, that, admit this be the true God whom we worship, and that be his Word, yet why may not the popish religion be the right? They have the same God, the same Christ, the same Word; they only interpret it one way, we another. This hath sometimes stuck with me, and more it would but the vain fooleries that are in their religion, together with their lying miracles and cruel persecutions of the saints, which admit were they as they term them, yet not so to be dealt withal. The consideration of these things and many the like would soon turn me to my own religion again.

But some new troubles I have had since the world has been filled with blasphemy and sectaries, and some who have been accounted sincere Christians have been carried away with them, that sometimes I have said, Is there faith upon the earth? and I have not known what to think. But then I have remembered the words of Christ that so it must be, and that, if it were possible, the very elect should be deceived. "Behold," saith our Saviour, "I have told you before." That hath stayed my heart, and I can now say, "Return, O my soul, to thy rest. Upon this rock Christ Jesus will I build my faith; and if I perish, I perish." But I know all the powers of hell shall never prevail against it. I know whom I have trusted, and whom I have believed, and that he is able to keep that I have committed to his charge.

Now to the King immortal, eternal, and invisible, the only wise God, be honor and glory for ever and ever! Amen.

This was written in much sickness and weakness, and is very weakly and imperfectly done; but if you can pick any benefit out of it, it is the mark which I aimed at.

HERE FOLLOW SEVERAL OCCASIONAL MEDITATIONS.

By night, when others soundly slept
    And had at once both ease and rest,
My waking eyes were open kept,
    And so to lie I found it best.

I sought him whom my soul did love;
    With tears I sought him earnestly;
He bowed his ear down from above.
    In vain I did not seek or cry.

My hungry soul he filled with good;
  He in his bottle put my tears;
My smarting wounds washed in his blood,
  And banished thence my doubts and fears.

What to my Saviour shall I give
  Who freely hath done this for me?
I 'll serve him here whilst I shall live,
  And love him to eternity.

FOR DELIVERANCE FROM A FEVER.

When sorrows had begirt me round,
  And pains within and out,
When in my flesh no part was found,
  Then didst thou rid me out.

My burning flesh in sweat did boil,
  My aching head did break;
From side to side for ease I toil,
  So faint I could not speak.

Beclouded was my soul with fear
  Of thy displeasure sore,
Nor could I read my evidence
  Which oft I read before.

"Hide not thy face from me," I cried;
  "From burnings keep my soul.
Thou knowest my heart, and hast me tried;
  I on thy mercies roll."

"Oh, heal my soul," thou knowest I said,
  "Though flesh consume to naught.
What though in dust it shall be laid?
  To glory it shall be brought."

Thou heardest, thy rod thou didst remove,
  And spared my body frail;
Thou showedst to me thy tender love,
  My heart no more might quail.

Oh, praises to my mighty God!
  Praise to my Lord, I say,
Who hath redeemed my soul from pit.
  Praises to him for aye!

FROM ANOTHER SORE FIT.

In my distress I sought the Lord,
  When naught on earth could comfort give;
And when my soul these things abhorred,
  Then, Lord, thou saidst unto me, Live.

Thou knowest the sorrows that I felt,
  My plaints and groans were heard of thee,
And how in sweat I seemed to melt;
  Thou helpedst and thou regardedst me.

My wasted flesh thou didst restore,
  My feeble loins didst gird with strength;
Yea, when I was most low and poor
  I said, "I shall praise thee at length."

What shall I render to my God
 For all his bounty showed to me —
E'en for his mercies in his rod,
 Where pity most of all I see?

My heart I wholly give to thee;
 Oh, make it fruitful, faithful, Lord!
My life shall dedicated be
 To praise in thought, in deed, in word.

Thou knowest no life I did require
 Longer than still thy name to praise,
Nor aught on earth worthy desire
 In drawing out these wretched days.

Thy name and praise to celebrate,
 O Lord, for aye is my request.
Oh, grant I do it in this state,
 And then with thee, which is the best.

DELIVERANCE FROM A FIT OF FAINTING.

Worthy art thou, O Lord of praise!
 But, ah! it 's not in me;
My sinking heart I pray thee raise,
 So shall I give it thee.

"My life as spider's web 's cut off!"
 Thus, fainting, have I said;
"And living man no more shall see,
 But be in silence laid."

My feeble spirit thou didst revive,
   My doubting thou didst chide;
And though as dead, madest me alive,
   I here a while might abide.

Why should I live but to thy praise?
   My life is hid with thee.
O Lord, no longer be my days
   Than I may fruitful be.

MEDITATIONS WHEN MY SOUL HATH BEEN REFRESHED WITH THE CONSOLATIONS WHICH THE WORLD KNOWS NOT.

Lord, why should I doubt any more when thou hast given me such assured pledges of thy love? First, thou art my Creator, I thy creature; thou my Master, I thy servant. But hence arises not my comfort. Thou art my Father, I thy child: "Ye shall be my sons and daughters," saith the Lord Almighty. Christ is my brother: "I ascend unto my father and your father, unto my God and your God." But lest this should not be enough, "Thy maker is thy husband." Nay, more, "I am a member of his body; he, my head." Such privileges, had not the word of truth made them known, who or where is the man that durst in his heart have presumed to have thought it? So wonderful are these thoughts that my spirit fails in me at the consideration thereof; and I am confounded to think that God, who hath done so much for me, should have so little from me. But this is my comfort: when I come into Heaven, I shall understand perfectly what he hath done for me, and then shall I be able to praise him as I ought. Lord, having this hope, let me purify myself as thou art pure, and let me be no more afraid of death, but even desire to be dissolved and be with thee, which is best of all.

July 8, 1656.

I had a sore fit of fainting, which lasted two or three days, but not in that extremity which at first it took me; and so much the sorer it was to me because my dear husband was from home, who is my chiefest comforter on earth. But my God, who never failed me, was not absent, but helped me, and graciously manifested his love to me, which I dare not pass by without remembrance, that it may be a support to me when I shall have occasion to read this hereafter, and to others that shall read it when I shall possess that I now hope for, that so they may be encouraged to trust in him who is the only portion of his servants.

O Lord, let me never forget thy goodness, nor question thy faithfulness to me; for thou art my God. Thou hast said, and shall not I believe it?

Thou hast given me a pledge of that inheritance thou hast promised to bestow upon me. Oh, never let Satan prevail against me, but strengthen my faith in thee till I shall attain the end of my hopes, even the salvation of my soul. Come, Lord Jesus; come quickly.

>What God is like to him I serve?
> What Saviour like to mine?
>Oh, never let me from thee swerve,
> For truly I am thine.
>
>My thankful mouth shall speak thy praise,
> My tongue shall talk of thee;
>On high my heart oh do thou raise,
> For what thou hast done for me.
>
>Go, worldlings, to your vanities,
> And heathen to your gods;
>Let them help in adversities,
> And sanctify their rods.
>
>My God he is not like to yours,
> Yourselves shall judges be;
>I find his love, I know his power,
> A succorer of me.
>
>He is not man that he should lie,
> Nor son of man to unsay;

His word he plighted hath on high,
  And I shall live for aye.

And for his sake that faithful is,
  That died but now doth live,
The first and last, that lives for aye,
  Me lasting life shall give.

———

My soul, rejoice thou in thy God;
  Boast of him all the day;
Walk in his law, and kiss his rod;
  Cleave close to him alway.

What though thy outward man decay,
  Thy inward shall wax strong;
Thy body vile it shall be changed,
  And glorious made ere long.

With angel's wings thy soul shall mount
  To bliss unseen by eye,
And drink at unexhausted fount
  Of joy unto eternity.

Thy tears shall all be driéd up,
  Thy sorrows all shall flee;
Thy sins shall ne'er be summoned up,
  Nor come in memory.

Then shall I know what thou hast done
  For me, unworthy me,

And praise thee shall e'en as I ought
  For wonders that I see.

Base world, I trample on thy face;
  Thy glory I despise;
No gain I find in aught below,
  For God hath made me wise.

Come, Jesus, quickly! Blessed Lord,
  Thy face when shall I see?
Oh, let me count each hour a day
  Till I dissolvéd be.

August 28, 1656.

After much weakness and sickness, when my spirits were worn out, and many times my faith weak likewise, the Lord was pleased to uphold my drooping heart, and to manifest his love to me. And this is that which stays my soul that this condition that I am in is the best for me, for God doth not afflict willingly, nor take delight in grieving the children of men. He hath no benefit by my adversity, nor is he the better for my prosperity; but he doth it for my advantage, and that I may be a gainer by it. And if he knows that weakness and a frail body is the best to make me a vessel fit for his use, why should I not bear it, not only willingly but joyfully? The Lord knows I dare not desire that health that sometimes I have had, lest my heart should be drawn from him, and set upon the world.

Now I can wait, looking every day when my Saviour shall call for me. Lord, grant that while I live I may do that service I am able in this frail body, and be in continual expectation of my change. And let me never forget thy great love to my soul so lately expressed, when I could lie down and bequeath my soul to thee, and death seemed no terrible thing. Oh, let me ever see thee that art invisible, and I shall not be unwilling to come, though by so rough a messenger.

May 11, 1657.

I had a sore sickness, and weakness took hold of me, which hath by fits lasted all this spring till this 11th May. Yet hath my God given me many a respite, and some ability to perform the duties I owe to him, and the work of my family.

Many a refreshment have I found in this my weary pilgrimage, and in this valley of Baca many pools of water. That which now I chiefly labor for is a contented, thankful heart under my affliction and weakness, seeing it is the will of God it should be thus. Who am I that I should repine at his pleasure, especially seeing it is for my spiritual advantage? For I hope my soul shall flourish while my body decays, and the weakness of this outward man shall be a means to strengthen my inner man.

"Yet a little while, and he that shall come will come, and will not tarry."

May 13, 1657.

> As spring the winter doth succeed,
>   And leaves the naked trees do dress,
> The earth all black is clothed in green,
>   At sunshine each their joy express.
>
> My sun's returned with healing wings,
>   My soul and body do rejoice;
> My heart exults, and praises sings
>   To him that heard my wailing voice.
>
> My winter's past, my storms are gone,
>   And former clouds seem now all fled;
> But if they must eclipse again
>   I'll run where I was succoréd.
>
> I have a shelter from the storm,
>   A shadow from the fainting heat;
> I have access unto his throne
>   Who is a God so wondrous great.
>
> Oh, thou hast made my pilgrimage
>   Thus pleasant, fair, and good;
> Blessed me in youth and elder age;
>   My Baca made a springing flood.

> I studious am what I shall do
>   To show my duty with delight;
> All I can give is but thine own,
>   And at the most a simple mite.

September 30, 1657.

It pleased God to visit me with my old distemper of weakness and fainting, but not in that sore manner sometimes he hath. I desire not only willingly, but thankfully, to submit to him, for I trust it is out of his abundant love to my straying soul which in prosperity is too much in love with the world. I have found by experience I can no more live without correction than without food. Lord, with thy correction give instruction and amendment, and then thy strokes shall be welcome. I have not been refined in the furnace of affliction as some have been, but have rather been preserved with sugar than brine; yet will he preserve me to his heavenly kingdom.

Thus, dear children, have ye seen the many sicknesses and weaknesses that I have passed through to the end that, if you meet with the like, you may have recourse to the same God who hath heard and delivered me, and will do the like for you if you trust in him. And when he shall deliver you out of distress, forget not to give him thanks, but walk more closely with him than before. This is the desire of your loving mother, A. B.

UPON MY SON SAMUEL HIS GOING FOR ENGLAND, NOVEMBER 6, 1657.

Thou mighty God of sea and land,
I here resign into thy hand
The son of prayers, of vows, of tears,
The child I stayed for many years.
Thou heardest me then, and gavest him me;
Hear me again: I give him thee.
He 's mine, but more, O Lord, thine own,
For sure thy grace on him is shown.
No friend I have like thee to trust,
For mortal helps are brittle dust.
Preserve, O Lord, from storms and wreck,
Protect him there, and bring him back;
And if thou shalt spare me a space,
That I again may see his face,
Then shall I celebrate thy praise,
And bless thee for it all my days.
If otherwise I go to rest,
Thy will be done, for that is best;
Persuade my heart I shall him see
For ever happified with thee.

May 11, 1661.

It hath pleased God to give me a long time of respite for these four years that I have had no great fit of sickness; but this year, from the middle of January till May, I have been by fits very ill and weak. The first of this month I had a fever seated upon me which indeed was the longest and sorest that ever I had, lasting four days; and the weather being very hot made it the more tedious. But it pleased the Lord to support my heart in his goodness, and to hear my prayers, and to deliver me out of adversity. But, alas! I cannot render unto the Lord according to all his lovingkindness, nor take the cup of salvation with thanksgiving as I ought to do. Lord, thou that knowest all things knowest that I desire to testify my thankfulness not only in word but in deed, that my conversation may speak that thy vows are upon me.

    My thankful heart with glorying tongue
        Shall celebrate thy name
    Who hath restored, redeemed, recured,
        From sickness, death, and pain.

    I cried, "Thou seemest to make some stay!"
        I sought more earnestly;
    And in due time thou succoredst me,
        And sentest me help from high.

    Lord, whilst my fleeting time shall last
        Thy goodness let me tell,
    And new experience I have gained
        My future doubts repel.

An humble, faithful life, O Lord,
    For ever let me walk;
Let my obedience testify
    My praise lies not in talk.

Accept, O Lord, my simple mite,
    For more I cannot give;
What thou bestowest I shall restore,
    For of thine alms I live.

FOR THE RESTORATION OF MY DEAR HUSBAND FROM A BURNING AGUE, JUNE, 1661.

When fears and sorrows me beset,
    Then didst thou rid me out;
When heart did faint and spirits quail,
    Thou comfortedst me about.

Thou raisedst him up I feared to lose,
    Regavest me him again;
Distempers thou didst chase away,
    With strength didst him sustain.

My thankful heart, with pen record
    The goodness of thy God;
Let thy obedience testify
    He taught thee by his rod,

And with his staff did thee support,
    That thou by both mayst learn,

And 'twixt the good and evil way
  At last thou mightest discern.

Praises to him who hath not left
  My soul as destitute,
Nor turned his ear away from me,
  But granted hath my suit.

UPON MY DAUGHTER HANNAH WIGGIN HER RECOVERY FROM A DANGEROUS FEVER.

Blest be thy name, who didst restore
  To health my daughter dear
When death did seem e'en to approach
  And life was ended near.

Grant she remember what thou hast done,
  And celebrate thy praise,
And let her conversation say
  She loves thee all thy days.

ON MY SON'S RETURN OUT OF ENGLAND, JULY 17, 1661.

All praise to him who hath now turned
  My fears to joys, my sighs to song,
My tears to smiles, my sad to glad:
  He's come for whom I waited long.

Thou didst preserve him as he went,
  In raging storms didst safely keep;
Didst that ship bring to quiet port —
  The other sank low in the deep.

From dangers great thou didst him free
  Of pirates who were near at hand,
And orderedst so the adverse wind
  That he before them got to land.

In country strange thou didst provide,
  And friends raised him in every place,
And courtesies of sundry sorts
  From such as 'fore ne'er saw his face.

In sickness when he lay full sore,
  His help and his physician wert;
When royal ones that time did die,
  Thou healedst his flesh, and cheered his heart.

From troubles and encumbers thou
  Without — all fraud — didst set him free,
That without scandal he might come
  To the land of his nativity;

On eagle's wings him hither brought
  Through want and dangers manifold,
And thus hath granted my request
  That I thy mercies might behold.

Oh, help me pay my vows, O Lord!
  That ever I may thankful be,

And may put him in mind of what
Thou hast done for him, and so for me.

In both our hearts erect a frame
Of duty and of thankfulness,
That all thy favors great received
Our upright walking may express.

O Lord, grant that I may never forget thy loving-kindness in this particular, and how graciously thou hast answered my desires.

UPON MY DEAR AND LOVING HUSBAND HIS GOING INTO ENGLAND, JANUARY 16, 1661.

O thou Most High, who rulest all,
  And hearest the prayers of thine,
Oh, hearken, Lord, unto my suit,
  And my petition sign.

Into thy everlasting arms
  Of mercy I commend
Thy servant, Lord; keep and preserve
  My husband, my dear friend.

At thy command, O Lord, he went,
  Nor naught could keep him back.
Then let thy promise joy his heart;
  Oh, help, and be not slack.

Uphold my heart in thee, O God,
  Thou art my strength and stay;
Thou seest how weak and frail I am,
  Hide not thy face away.

I, in obedience to thy will,
  Thou knowest did submit;
It was my duty so to do;
  O Lord, accept of it.

Unthankfulness for mercies past
  Impute thou not to me;
O Lord, thou knowest my weak desire
  Was to sing praise to thee.

Lord, be thou pilot to the ship,
  And send them prosperous gales;
In storms and sickness, Lord, preserve —
  Thy goodness never fails.

Unto thy work he hath in hand,
  Lord, grant thou good success
And favor in their eyes to whom
  He shall make his address.

Remember, Lord, thy folk whom thou
  To wilderness hast brought;
Let not thine own inheritance
  Be sold away for naught,

But tokens of thy favor give —
  With joy send back my dear,

That I, and all thy servants, may
    Rejoice with heavenly cheer.

Lord, let my eyes see once again
    Him whom thou gavest me,
That we together may sing praise
    For ever unto thee;

And the remainder of our days
    Shall consecrated be
With an engagéd heart to sing
    All praises unto thee.

IN MY SOLITARY HOURS IN MY DEAR HUSBAND HIS ABSENCE.

O Lord, thou hearest my daily moan,
    And seest my dropping tears;
My troubles all are thee before,
    My longings and my fears.

Thou hitherto hast been my God,
    Thy help my soul hath found;
Though loss and sickness me assailed,
    Through thee I've kept my ground.

And thy abode thou hast made with me;
    With thee my soul can talk
In secret places, thee I find
    Where I do kneel or walk.

Though husband dear be from me gone,
    Whom I do love so well,
I have a more beloved one
    Whose comforts far excel.

Oh, stay my heart on thee, my God,
    Uphold my fainting soul;
And when I know not what to do
    I'll on thy mercies roll.

My weakness thou dost know full well
    Of body and of mind;
I in this world no comfort have
    But what from thee I find.

Though children thou hast given me,
    And friends I have also,
Yet if I see thee not through them
    They are no joy, but woe.

Oh, shine upon me, blessed Lord,
    E'en for my Saviour's sake;
In thee alone is more than all,
    And there content I'll take.

Oh, hear me, Lord, in this request,
    As thou before hast done —
Bring back my husband, I beseech,
    As thou didst once my son.

So shall I celebrate thy praise
    E'en while my days shall last,

And talk to my beloved one
  Of all thy goodness past.

So both of us thy kindness, Lord,
  With praises shall recount,
And serve thee better than before
  Whose blessings thus surmount.

But give me, Lord, a better heart;
  Then better shall I be
To pay the vows which I do owe
  For ever unto thee.

Unless thou help, what can I do
  But still my frailty show?
If thou assist me, Lord, I shall
  Return thee what I owe.

IN THANKFUL ACKNOWLEDGMENT FOR THE LETTERS I RECEIVED FROM MY HUSBAND OUT OF ENGLAND.

O thou that hearest the prayers of thine,
And 'mongst them hast regarded mine,
Hast heard my cries, and seen my tears,
Hast known my doubts and all my fears,

Thou hast relieved my fainting heart,
Nor paid me after my desert;
Thou hast to shore him safely brought
For whom I thee so oft besought.

Thou wast the pilot to the ship,
And raised him up when he was sick;
And hope thou hast given of good success
In this his business and address,

And that thou wilt return him back
Whose presence I so much do lack.
For all these mercies I thee praise,
And so desire e'en all my days.

IN THANKFUL REMEMBRANCE FOR MY DEAR HUSBAND'S SAFE ARRIVAL, SEPTEMBER 3, 1662.

What shall I render to thy name,
  Or how thy praises speak;
My thanks how shall I testify?
  O Lord, thou knowest I'm weak.

I owe so much, so little can
  Return unto thy name,
Confusion seizes on my soul,
  And I am filled with shame.

Oh, thou that hearest prayers, Lord,
  To thee shall come all flesh;
Thou hast me heard and answeréd,
  My plaints have had access.

What did I ask for but thou gavest?
  What could I more desire

Occasional Meditations

But thankfulness e'en all my days?
  I humbly this require.

Thy mercies, Lord, have been so great,
  In number numberless,
Impossible for to recount
  Or any way express.

Oh, help thy saints that sought thy face
  To return unto thee praise,
And walk before thee as they ought
  In strict and upright ways.

["This was the last thing written in that book by my dear and honored mother."— Note by Simon Bradstreet, Jr.]

["Here follow some verses upon the burning of our house, July 10th, 1666. Copied out of a loose paper."— Note by Simon Bradstreet, Jr.]

In silent night, when rest I took,
For sorrow near I did not look.
I wakened was with thundering noise
And piteous shrieks of dreadful voice.
That fearful sound of "Fire!" and "Fire!"
Let no man know, is my desire.

I, starting up, the light did spy,
And to my God my heart did cry

To strengthen me in my distress,
And not to leave me succorless;
Then coming out, beheld apace
The flame consume my dwelling-place.

And when I could no longer look
I blest his name that gave and took,
That laid my goods now in the dust;
Yea, so it was, and so 't was just —
It was his own; it was not mine.
Far be it that I should repine.

He might of all justly bereft,
But yet sufficient for us left.
When by the ruins oft I passed
My sorrowing eyes aside did cast,
And here and there the places spy
Where oft I sat, and long did lie.

Here stood that trunk, and there that chest;
There lay that store I counted best;
My pleasant things in ashes lie,
And them behold no more shall I.
Under thy roof no guest shall sit,
Nor at thy table eat a bit;

No pleasant tale shall e'er be told,
Nor things recounted done of old;
No candle e'er shall shine in thee,
Nor bridegroom's voice e'er heard shall be.

In silence ever shalt thou lie.
Adieu, adieu; all 's vanity.

Then straight I 'gan my heart to chide:
And did thy wealth on earth abide?
Didst fix thy hope on mouldering dust,
The arm of flesh didst make thy trust?
Raise up thy thoughts above the sky,
That dunghill mists away may fly.

Thou hast an house on high erect;
Framed by that mighty Architect,
With glory richly furnished,
Stands permanent though this be fled.
It 's purchaséd, and paid for, too,
By Him who hath enough to do —

A price so vast as is unknown,
Yet, by his gift, is made thine own.
There 's wealth enough; I need no more.
Farewell, my pelf; farewell, my store;
The world no longer let me love.
My hope and treasure lie above.

[Another loose paper.]

As weary pilgrim, now at rest,
   Hugs with delight his silent nest,
His wasted limbs now lie full soft
   That miry steps have trodden oft;
Blesses himself to think upon
   His dangers past and travails done;
The burning sun no more shall heat,
   Nor stormy rains on him shall beat;
The briars and thorns no more shall scratch,
   Nor hungry wolves at him shall catch;
He erring paths no more shall tread,
   Nor wild fruits eat, instead of bread;
For waters cold he doth not long,
   For thirst no more shall parch his tongue;
No rugged stones his feet shall gall,
   Nor stumps nor rocks cause him to fall;
All cares and fears he bids farewell,
   And means in safety now to dwell —
A pilgrim I on earth, perplexed
   With sins, with cares and sorrows vexed,
By age and pains brought to decay,
   And my clay house mouldering away,
Oh, how I long to be at rest,
   And soar on high among the blest!
This body shall in silence sleep,
   Mine eyes no more shall ever weep;

No fainting fits shall me assail,
  Nor grinding pains my body frail,
With cares and fears ne'er cumbered be,
  Nor losses know, nor sorrows see.
What though my flesh shall there consume?
  It is the bed Christ did perfume;
And when a few years shall be gone
  This mortal shall be clothed upon.
A corrupt carcass down it lies,
  A glorious body it shall rise;
In weakness and dishonor sown,
  In power 't is raised by Christ alone.
Then soul and body shall unite,
  And of their maker have the sight;
Such lasting joys shall there behold
  As ear ne'er heard nor tongue e'er told.
Lord, make me ready for that day!
  Then come, dear bridegroom, come away.

August 31, 1669.